We Got It From THEM!

How Humanity Received ET Technology

By Dr. Gregory Rogers

Former USAF/NASA Flight Surgeon

We Got It From THEM!

How Humanity Received ET Technology

By Dr. Gregory Rogers

Published in the United States of America

ISBN# 979-8-9887888-8-1

Publisher:

Un-X Media
PO Box 1166
Independence, MO 64050
www.unxmedia.com

UNXMEDIA

PUBLISHING

Dedication

Dedicated to my wife Judy

and

Julie and Adin, Atira, Nick, Adin

Greg and Jessica, Hudson, Hannah, Caroline

Cindy and Daniel, William

Testimonial from Thomas De La Garza, Ph.D.

Thomas De La Garza, Ph.D. is a retired military officer with over 20 years of experience in multi-domain and special operations. He holds the privilege of serving in multiple regional combatant commands, NORAD, United States Strategic Command, and United States Space Command. Since childhood he has held great interest in UFOs and has been witness to two sightings, thankfully while in the company of friends.

Dr. Gregory Rogers uses his unique blend of medical knowledge and aviation expertise to provide crucial context regarding historical UFO and UAP phenomenon. He deftly relates these incidents to the extraordinary he was exposed to during his tenure at NASA. While Dr. Rogers is among a growing list of patriotic individuals willing to discuss government involvement in this sensitive area, few have his medical and scientific credentials. This book is a must read for those interested in better understanding UAP capabilities and their relationship with mainstream technologies.

~ Thomas De La Garza, Ph.D.

Testimonial from Mindy Tautfest

As the re-founder and CEO of the International UFO Bureau (IUFOB), my role has always centered on building credible, compassionate pathways to understanding the phenomenon. Because of my background developing investigative training, used by thousands across the field, and my experience as a trusted speaker and media contributor, I understand the importance of pairing technical accuracy with human-centered communication. This balanced approach has shaped IUFOB's mission to unify rigorous research with public engagement. It's also why Dr. Rogers was one of the first people invited to serve on our Board of Directors. His deep insight into the phenomenon, combined with his unwavering integrity, made him an essential voice in helping guide the future of our work. Our shared commitment to transparency, truth, and collaboration forged a strong professional bond from the very beginning.

That same clarity of vision and purpose shines through in Dr. Rogers' new book, a powerful and timely contribution to the field of Unidentified Anomalous Phenomena. He skillfully outlines the credible historical record of these encounters while emphasizing the growing need for a unified agency to lead this important research. Drawing from his unmatched understanding of aerospace systems and firsthand knowledge of long-standing efforts to reverse engineer non-human technology, Dr. Rogers delivers a narrative that is both authoritative and deeply revealing. His courage in bringing forward once-hidden truths challenges long-held institutional silence and empowers the public to be part of the conversation. There is strength in numbers, and as experts and citizens come together, we move closer to revealing the truth that belongs not to governments or select groups, but to all of humanity.

~Mindy Tautfest

Acknowledgements

I want to thank a number of people who have assisted me as I have revealed this story.

From the International UFO Bureau, I must thank Mindy and John Tautfest, Teri Lynge-Kehl, Robert Dio, Melissa Madrigal, Jonathan Dover, Michele Meiners, Jeremy Ray, Dr. Tom De La Garza, Dr. Ed Wolff, Dan Borneman

Special thanks to Apollo 11 astronaut Buzz Aldrin and Apollo 13 astronaut Jim Lovell, Jr.

I would also like to thank Linda Moulton Howe, Chris Lehto, Tyler Roberts, Josh Boswell, Jesse Michels, former Rear Admiral Tim Gallaudet, Martin Willis, Todd Curtis, and Kerry Cassidy

Table of Contents

Introduction

I am **Dr. Gregory Rogers**, former U.S. Air Force Chief of Aerospace Medicine for the 45th Space Wing. As such, I dealt with military and NASA operations at Patrick Air Force Base, Cape Canaveral Air Force Station, the Eastern Space and Missile Center, and the Eastern Missile Range. In conjunction to these duties, I was attached to NASA through the DDMS, whose name has been altered many times. The simplest of these variations was the Department of Defense Manned Spaceflight Support Office. Through DDMS, I supported 31 space shuttle missions. This included launch and landing operations, plus many exercises to function within the emergency contingency modes that would have been needed for astronaut rescues during such missions.

Unidentified Flying Objects, or UFOs, which have recently been known by the preferred name of Unidentified Aerial/Anomalous Phenomena, or UAPs, have long hovered at the crossroads of hard science, rampant speculation, and definitely an ongoing public fascination. From mere fleeting lights in the nighttime sky to an array of detailed encounters reported by pilots, military personnel, and civilians alike, UFOs have provoked both great wonder and immense skepticism.

While the term "UFO" simply denotes any aerial phenomenon not immediately identifiable, especially within the military tactical settings I have been familiar with through the years, its cultural connotations abound. For many people, the term itself has become toxic. This explains why the UAP replacement identification has become so popular as its substitute, particularly within the governmental and scientific communities. Many ardent believers suggest there are extraterrestrial visitors, secret advanced technologies, and a long history of global interactions. For non-believers, they lean toward simpler explanations, such as misinterpretations of natural events.

In this book, I want to go beyond the superficial reports as we explore what science can tell us about UAPs. We will study their amazing history, how they have been and are now investigated, what explanations have been proposed, and what burning questions surely remain unanswered.

I would tell you that I am an unbiased writer of this book, but I cannot say that. I believe that UFOs/UAPs are REAL, and this clearly means that I will be supporting my views in what you will read. With all I have done in my life, I feel a responsibility to utilize my experiences to guide my readers along a particular line of logic. This is partially due to my background as a physician.

Medical thought processes are outcome-related. I believe that if you have appendicitis, you need surgery. If you are having blood-clots in your brain causing a stroke, you need to receive a thrombolytic agent, as swiftly as possible. If you are hypothyroid, you need a thyroid supplement to balance your body functions. From my military background, I figure that if you pick up a live grenade, do not pull the pin unless you are ready to throw it in a combat situation. From a scientific background, I expect that bosons will always have whole number rotations based on the Planck Constant. I believe the Pauli Exclusion Principle dictates that no two fermions can occupy the same quantum state yet does not affect the activity of the bosons I mentioned earlier.

I believe these are truths. What I will similarly present in this book are also what I believe to be significant "truths" of nature.

Now, let us proceed.

My Story

For specific context as to why I believe what I believe, allow me to tell my
own story:

1. I am a God-fearing and loyal American who was extremely proud
   to previously serve this country as a field grade military officer and
   later as a Department of Defense civilian physician. I recently
   retired from the Defense Health Agency and have decided that I
   should now reveal information about a significant event that
   happened to me back in 1992.

2. At the time, I was a major in the U.S. Air Force and serving as the
   Chief of Aerospace Medicine for the 45th Medical Group, as part
   of the 45th Space Wing. We operated Patrick Air Force Base
   (AFB), the Cape Canaveral Air Force Station (CCAFS), and the
   Eastern Space and Missile Center. We likewise operated control of
   the Eastern Missile Range, as well as supporting operations for
   multiple tenant organizations. These would include AFTAC (the
   Air Force Technical Applications Center) and the U-2 spy planes
   of the 9th Strategic Reconnaissance Wing. We also supported
   NASA operations, both on the Cape Canaveral Air Force Station
   side and the Kennedy Space Center (KSC) side.

3. I usually assisted NASA space shuttle missions through the
   DDMS, or Department of Defense Manned Spaceflight Support
   Office. I similarly had the privilege of performing numerous air-
   sea rescues in the Atlantic Ocean aboard the HH-3 helicopters of

the 41st Air Rescue Squadron and the HH-60 helicopters of the 301st Rescue Squadron. I was awarded a Sikorski Rescue Award, a Humanitarian Service Medal, and an Air Medal due to the joint efforts to support the people of South Florida after the devastation caused by Hurricane Andrew. Being from Oklahoma, I had seen terrible tornado damage before. For Hurricane Andrew, however, the damage appeared to be from a tornado that was at least sixty miles wide. I was impressed by this storm, but for all of the wrong reasons.

4. I served as a rescue flight surgeon for the astronauts at KSC, or the Kennedy Space Center, during 31 space shuttle launches and then their subsequent landings at both Kennedy Space Center and at Edwards AFB in California, via the Support Operations Center at the CCAFS.

5. Starting with my arrival to Patrick Air Force Base in 1989, I received specialized training from the Department of Defense, including the Air Force Space Operations Course and also the STS (Shuttle) Orbiter Familiarization Course, both held at the Cape Canaveral Air Force Station. I was then certified as a NASA flight surgeon at Johnson Space Center, Texas. Similarly, I also supported NASA and Air Force launches of Titan, Atlas, and Delta vehicles operating from the Cape. There were additional military duties I performed that are not suitable for a discussion here. This period of my service to the U.S. Air Force and NASA included during the year of 1992.

6. I would like to specify that I assuredly believe that the universe was created by God, and that I feel all facts of our sciences unmistakably point to this being true. I believe that the so-called "Big Bang" and the moment in Genesis when God said "Let there be Light" each describe exactly the same event. As such, God has created an immensely complex universe that continues to unfold before our eyes. God has many mysteries that are far beyond our knowledge and experience. Since God created the universal characteristics of all physics, chemistry, and metallurgy, this is how we humans have been allowed to use our knowledge of these sciences to create chariots, cars, aircraft, and spacecraft. If we have done this on our planet, is it completely impossible for other beings to have created these same types of technological devices on their own planets?

7. We know from the Bible, and other forms of scriptures, that we humans are not His only creations. Many other beings have been repeatedly revealed to us from within our holiest writings. There are reported to be angels, seraphim, cherubim, the Heavenly hosts, demons, watchers, and the mysterious "sons of God." Are these that are listed the only other creatures within the entirety of our universe? I do not believe so.

8. We, as human beings, believe we are highly intelligent, and we truly have learned so much about this vast and magnificent creation. This is partially true. Nevertheless, there are far more truths that we do not comprehend than all that we do recognize through science. This is more noteworthy over just the last few

hundred years. I fully believe that if we could somehow see ahead just 200 years into our future, the level of our own human sciences available at that time would completely dwarf the level of physics and other sciences that we are so very proud to know today. There will be facts available then that we cannot conceive of today. We have so very much more truths to learn about this fantastic creation.

9. In other words, we humans are <u>NOT</u> nearly as smart as we think we are. There is far more within this universe that we do <u>not</u> know compared to what we fully understand in our times. These enigmatic truths could absolutely involve the presence of galactic or dimensional intelligent life forms other than what we know to exist on our small blue planet. We would be haughty indeed for us to believe that we are all there is for intelligent life within the entirety of the vast cosmos!

10. With this said, I will address the experience I witnessed on Cape Canaveral Air Force Station in 1992. As best as I recall, the event was in late spring of that year. What I witnessed was in no way expected, desired, nor selected by me. Instead, it was a matter of sheer happenstance, from my viewpoint. Had I been given a choice, I would have chosen NOT to have been a part of this experience. The sad truth is that it happened in a fashion that both surprised me and then made me quite angry afterward.

11. I had been to a facility on the Cape in my responsibility for oversight of some of the medical operations on the Cape, in my

role as Chief of Aerospace Medicine for the 45th Space Wing. The functional contractor was EG&G, which stands for Edgerton, Germeshausen, and Grier, Inc, which is a major national defense contractor. There were also a large number of other major DoD and space contracting organizations that I held a supporting role in dealing with, both on the Cape and over at Kennedy Space Center.

12. On this specific day, I had just finished being shown a new function within one of our many Cape buildings and my EG&G contractor guide had left the building as soon as we finished my introduction to the process. To enter that bay, I had been required to wear a hair cap, lab coat, gloves, and non-electrostatic booties. As we egressed that bay, I was slower than my guide in doffing my gear, so I was a minute or two behind him. As I entered the hallway to get to the exit, the escort had left but an unfamiliar major stopped me within this hallway. He told me he had something important that he wanted to show me. Having no idea what it might be, I then foolishly accepted his unanticipated invitation.

13. I need to explain that I and my flight medicine team performed annual physicals and routine medical care for a large range of pilots, aircrews, many support personnel within flight organizations, air traffic controllers, combat air controllers, missile crews, AFTAC personnel, DDMS personnel, and even U.S. Navy personnel operating at NOTU (Naval Ordnance and Test Unit) out of Port Canaveral Naval Station. We sometimes supported

transient NASA astronauts, visiting dignitaries, personnel who supported a unique State Department mission outside of U.S. territorial waters and airspace, and sundry other specialized functions.

14. Because of this, the major knew me even though I was unfamiliar with him, for a good reason. As is often the case with doctors, patients recall us far better than we recall them. I might see a fair number of sick call patients and up to eight flight or specialized physical patients in a single day. Because of the considerable number of those I had seen, patients would tend to run together for me for even a months' time. I certainly did not always remember patients I might see only once a year, or even only every other year. Still, since coming to the flight surgeons' office was so unique to them, they all knew me because there were only two full time flight surgeons present for all of the patients from all of the units I mentioned before.

15. When he approached me, this major stated he had something to show me that even I had never seen. I have presumed, since that time, that he said this because I had security clearances to go to most of the facilities on Cape Canaveral, often requiring security escorts, while most workers only had their approved clearances to go to the specific locations where they personally worked. The idea of this was to maintain security compartmentalization. This would limit the risk of people from one functional area learning about key classified tasks performed in other locations that they did not have either the proper clearance nor the actual need-to-

know for that information. Because I sometimes travelled all over the Cape, I saw much more of our facilities and operations than did the common workers.

16. This officer quickly directed me into a nearby office, then locked the door behind us. There were louvered blinds on the windows looking out into the hall, and he closed these. He then commented that he was going to show me something that would knock my socks off. He next sat down at a computer station and turned on the screen and computer. He took a minute or two to find the video he wanted, then had me sit down in his chair while he moved to one just beside me to my left. I had absolutely no idea what I was about to see and witness.

17. On the screen, there came a video that appeared to be a closed-circuit television feed from a fairly typical flight hangar. There were no markings of any kind on this video. It did NOT show any kind of classification indicators, location indicators, timeframe indicators for when the video was recorded, nor any other type of explanatory displays on the screen. However, what I saw on the screen was extremely unique to my experience.

18. What I saw was a white, completely smooth vehicle that appeared to be an actual flying saucer. It was constructed in the shape of a modified circular egg, being equal in length and shape from all sides, as I would learn. Its contour was totally smooth, with no lights, flight control surfaces, pitot tubes, propulsion systems, antennae, or other projections that might be seen on any typical

aircraft or spacecraft. While it was ovoid from side to side, it was about twenty feet in length, I would guess. It had a type of beam or equatorial region, with about 60 percent of the craft being above this region and only 40 percent of the craft being below this region. A shallow dome and upward mast was seen at its apex.

19. It was completely white, with no seams, rivets, or other clues to its construction. The only exceptions were a set of vertical and horizontal black rectangles that I assumed were markings for use with evaluating a test vehicle. The three vertical rectangles were only on the upper section of the vehicle, located at what I would call the 3 o'clock, 6 o'clock, and 9 o'clock positions. The four horizontal rectangles were along the beam, or equatorial positions, located from 12:30 to 2:30, then 3:30 to 5:30, then from 6:30 to 8:30, then 9:30 to 11:30 positions. When I initially saw the vehicle, the camera angle was from somewhat above the vehicle, looking down on it from the 3 o'clock position. I gained that rest of the information about these rectangles only as I witnessed this craft moving later on.

20. I could partially make out information at the 12 o'clock position, but did not get a full view of it until it initially rotated in the clockwise direction. I am calling this the 12 o'clock position because there, just above the equatorial region, was writing that said "U.S. Air Force" and just above that was the flight insignia for U.S. aircraft. This was the blue circle with a white star on it and red and white bars extending to the left and right of the blue margins.

21. As I mentioned, there was a shallow, blended dome to the apex of the vehicle, and a mast extended straight up from the vertex of this dome. The mast was then attached to what appeared to be three umbilical hoses that twirled around each other, and these extended from above the vehicle and upward, out of my sight.

22. Alongside the vehicle, to the lower left of the screen I saw two men who were in lab coats. I presumed these were engineers. To the right side, there were three technician-types walking around in full-body suits similar to what I would refer to today as Tyvek® suits. When a type of warning klaxon went off, all five of these individuals immediately moved out of my view from the screen I was seeing. It was the comparison of these people that caused me to estimate the craft to be about twenty feet in length.

23. I turned to the major and asked, "Who would build such a vehicle as that?" He said he could not tell me. I then asked, "Why would we design a craft configured like that?" His response was to say, "We got it from them!" He then used his thumb to point directly up, motioning this direction several times.

24. All of a sudden, there were what appeared to be visual and auditory signs of electromagnetic discharges occurring around the vehicle. I will not describe this otherwise, because even though there were no indications of the craft being classified, the fact that it was a flying saucer made me immediately expect that this surely was a classified craft. As such, I do not wish to disclose any information that might indicate its method of operations,

propulsion, or other characteristics that might still be used by it or even more advanced craft within the current inventory of the United States.

25. The craft rose from the concrete floor so smoothly that if I had not been closely watching, I would not have recognized the motion. It then smoothly rotated one complete cycle going clockwise, then did the same rotating counterclockwise. It was once more in the same position it had been when it initially rose from the floor. It then moved right and left, then forward and backward, which I assumed was simply testing of the controls for its motion.

26. All of these actions seemed routine, noting that it was a flying saucer. However, what it did next surprised me greatly. The vehicle rose from the front of the craft, tilting to a 45-degree forward angle-of-attack, but absolutely maintained its stationary position over the floor from where it had risen. I gave some exclamation and the major commented that he thought this would blow me away.

27. Just then, there were knocks on the door and voices asking why the door was locked. The major immediately turned off the computer console and screen, then looked at me very seriously. He passionately whispered, "Don't tell anyone I showed this to you!"

28. He then unlocked the door and three officers came in. One was a lieutenant colonel, who again asked why the door had been shut

and locked. The major stated that he had wanted to show me a skin lesion, so he had locked the door for the sake of his privacy.

29. At this point, my mind was racing. If the video had shown any type of classification markings, I would have definitely had to reveal that I had been shown a classified video in an improper setting. I also considered that someone had to have shown him the video, properly or not, or the major would not have known to look it up to show me. If any of these officers had shown it to him inappropriately, then they might also be guilty of the same error.

30. However, this was 1992 and for me to say that I had just seen a flying saucer video would have been a huge deal. If I made any comment to these four men, I would then be required to report these men to their commander and their chain-of-command. Then I would also be required to report this to our security personnel, not to mention those officers within my own chain-of-command. I could not imagine having to tell my colonel I had just seen a flying saucer and then reported a security violation. This seemed to me to be a career-deciding outcome.

31. Quickly making my choice to avoid commenting on that odd video, therefore, I remarked that the major's skin lesion was not cancerous and stated that I had to return to my clinic. I rapidly left that building and never returned. During my drive back to Patrick from the Cape, throughout this thirty-minute drive, I was angered and concerned about what I had witnessed. I finally conclusively

decided that I could not report this event, to anyone, due to the possible range of unknown repercussions I might personally receive as the consequence of making such a bizarre report. Therefore, for more than 30 years, I kept silent about this entire adventure. I did not even tell my wife, who was the closest person to me, until about fifteen years later.

32. Having had this experience, the later whistleblower videos, and recent testimony before Congress has captured my attention. Late in 2023, I decided to try something, so I requested from the commander of the Navy's Professional Development Symposium, or PDS, the permission to speak about the Navy's F/A-18 videos of UAPs during the lecture I was planning to provide for April of 2024. For several years in a row, I had given a series of specialized presentations to the PDS. I had previously spoken about an introduction to DoD explosive munitions for medical personnel, keys for monitoring and preventing the development of anemia in TNT and Tritonal production workers; hazardous noise evaluation and abatement methods; and even the recognition and treatment of worldwide entomology threats to our military personnel deploying to dangerous places around the planet.

33. After a couple of emails back and forth, I was given permission to explicitly discuss UAPs during my April 2024 PDS lecture. It was officially titled: HUMAN FACTORS IN IDENTIFICATION OF AIRCRAFT, MILITARY DRONES, AND OTHER AERIAL PHENOMENA. Having sent a copy of my slide presentation to the PDS personnel, it was approved for me to

deliver. To make certain there were no other conflicts, I sent my slide presentation to the Public Affairs officers at both Fort Sill, Oklahoma, and at the McAlester Army Ammunition Plant, where I worked at that time. I received formal written approval from both sets of officers.

34. Therefore, I gave my presentation in April of 2024. For the final 15 minutes or so, I discussed the Navy's release of official footage of UAPs as was recorded by naval aviators that included visual and auditory confirmation that F/A-18 Super Hornet crews had witnessed, recorded, and reported. The UAPs they had intercepted were not at all ordinary aircraft. Then this evidence had been declassified by the DoD and fully released to the public. I discussed my personal opinion that these vehicles exhibited flight characteristics that could not, at all, be duplicated by any human technology I knew about from either the DoD or NASA.

35. The presentation I gave virtually to military members all around the world, was very well received and I even received a letter of congratulations for the presentation I gave that day. It was from the same commander of the PDS who had approved my presentation.

36. With this in mind, now that I have fully retired from the governmental Defense Health Agency, I have decided to release this surprising information about what I witnessed at the CCAFS back in 1992. I have seen and heard the recent presentations given to members of Congress about UAPs. I recently have heard a

number of the members of Congress express their desire to hear more of the experiences of any additional whistleblowers. They have certainly encouraged any other DoD employees to testify concerning their own information about UAPs. I have specifically heard testimony of whistleblowers who have firmly declared that the government has attempted to reverse engineer technology based on UAPs. Therefore, I felt I could provide confirmation about this and so I am presenting my story.

37. For one thing, someone had to design and manufacture the vehicle I saw on that video. Someone, somewhere, had to pay for the research and development of this vehicle. I believe that this monetary expenditure should have been identified to Congress, since the Constitution declares that they hold the power of the purse over all military spending. While I do not know if that video was from Cape Canaveral Air Force Station, from the legendary Lockheed Skunk Works in California, from Area 51 in Nevada, from Wright-Patterson AFB in Ohio, or from any other DoD or military contractor location, the technology I witnessed that day was quite unusual to my awareness.

   a. First, it had no identifiable flight control surfaces or visible means of propulsion.

   b. Second, it was able to maintain a tilted flight configuration while holding a stable 45-degree forward angle-of-attack. I have personally piloted both fixed-wing and rotary-winged aircraft and none of them ever had the capability to maintain stability in the manner that this craft demonstrated.

c. Thirdly, with the exception of the lecture I personally gave in April of 2024, there have not been any official explanations, that I know of, that were given to the members of the Department of Defense about UAPs or vehicles such as I have described. I feel this is a huge mistake, because I know exactly what I saw that day, and it was in no fashion a conventional flying vehicle. For other military members, when they see similar advanced technology, they deserve to be listened to as they speak about their events.

d. I do not believe that Congress or the American public have officially received any approved description of any DoD or NASA aircraft or spacecraft that could perform as I saw on that day back in 1992. In fact, I have heard repeated "official" reports that no such vehicle as I witnessed has now or ever belonged to the American military. This statement, however, does not necessarily include possible research and development projects operated by the many contracting organizations functioning on behalf of the DoD. The recent revelation of a large number of governmental declassified reports about UAPs have been the only source of information that even remotely resembled what I saw.

e.   Other whistleblowers that have learned about UAPs have given their own testimony to Congress, so I am simply adding my voice to theirs in confirming their prior reports. As they have led the way by speaking out before me, I hope that my report may stimulate others to reveal whatever they have similarly witnessed.

In conclusion, I absolutely resent being a part of the events of that day. I surely had no desire to participate in those events. Still, I definitely believe that we have learned much as humans, but that we must remain open to new truths once they have been repeatedly demonstrated to us. To my mind, we are beyond the point of deciding whether UAPs are real, because the evidence today is, to my viewpoint, clear and unambiguous. What is left for us to decide is whether we will accept this fact, and try to cope with this reality. Otherwise, we may act in the manners of the foolish errors of the past, such as denying that the sun is the center of the solar system, that stars and planets only exist throughout our galaxy, and denying the rest of our amazing cosmos.

I firmly believe that the craft I unsuspectingly observed that day was built by mankind but doubtlessly was designed based on the reverse engineering of some craft that had demonstrated capabilities that are beyond our traditional technology. My account, I believe, is merely one additional indication of this indisputable fact. However, let me make one point very clearly:

<u>YOU CAN'T REVERSE ENGINEER A UNIQUE SPACECRAFT THAT NEVER EXISTED!</u>

I sincerely pray that all of mankind may see that we live on but a tiny speck of land and sea on a tiny blue planet within a great and fabulous universe. Moreover, we are not alone in this amazing universe. Realizing this, may we citizens of this world finally learn to treat our human neighbours with kindness and peace, for this is the only planet that we possess. Our national and religious differences are far less than what we might share with extraterrestrial visitors already amongst us. May God bless us all and bless the United States of America!

Chapter 1:

A Brief History of The Most Discussed UFO Sightings

Reports of strange objects in the sky are assuredly not unique to the modern era. Historical records, going back many centuries, document similar unusual aerial phenomena. These would include fiery wheels, mysterious lights, and even flying "shields" and "crosses." During World War II, aircrews across Europe identified strange orbs of light that performed amazing actions. These were known as "Foo fighters" and still remain unexplained since that time. As a matter of fact, my own father-in-law saw these Foo fighters while flying in B-17s over Germany. However, the modern UFO era began in earnest after World War II. In 1947 private pilot Kenneth Arnold reported nine crescent-shaped objects flying at incredible speed near Mount Rainier, coining the term "flying saucers." We will see more of his adventure later in our story.

The 1950s and 1960s saw a surge in sightings, with thousands of cases reported across the nation and even worldwide. High-profile incidents abound, such as the well-known Roswell crash in New Mexico in 1947, the notable 1961 Betty and Barney Hill abduction, the strange Kecksburg UFO Incident of 1965, and the amazing story of what all happened at the Malmstrom Air Force Base in Montana during March of 1967.

These have jointly fueled public imagination and governmental concern. We will later take a look into some of the important episodes in more recent times. Although the details of many of these may be familiar to many, I want to look further into each of these incidents.

❖ The Roswell Crash: A Detailed Account

Few episodes in American history have captured the public imagination quite like the 1947 Roswell incident. Often cited as the quintessential UFO event, the so-called "Roswell crash" blends elements of mystery, government secrecy, and extraterrestrial speculation. Since then, the strange story of what exactly happened near Roswell, New Mexico, has spawned countless books, documentaries, debates, and even a subculture of dedicated UFO researchers. This can be easily seen by the enormous number of t-shirts, hats, coffee mugs, and other merchandise snatched up by those who find the story so interesting. Let us explore the details of this Roswell crash, from the first reports to the enduring debates and cultural legacy that remain so persistent.

The Discovery in the Desert

In early July 1947, a rancher named William "Mac" Brazel discovered unusual debris scattered across his ranch, located about 75 miles north of Roswell, New Mexico. The debris included metallic rods, foil-like material, and pieces that some described as resembling parchment or unusual rubber. Brazel, puzzled by the find, gathered some of the strange metallic materials and brought it to the attention of the local authorities. He initially contacted Sheriff George Wilcox at the nearby town of Corona, who in turn notified personnel at the nearby Roswell Army Air Field (RAAF).

This is something that should not be overlooked. At RAAF, there resided the renowned 509th Bomb Group. These B-29 bombers and aircrews in 1945 were the ones that had dropped the nuclear bombs on Hiroshima and Nagasaki less than two years before. This was, to that moment, the only operational nuclear bombing group in history!

On July 8, 1947, the RAAF officer Major Jesse Marcel Sr. went with Brazel to the site and helped to recover debris. The result of all this activity was that 1st Lt. Walter Haut issued a press release stating that they had actually recovered a "flying disc" from a ranch near Roswell. This announcement was published in the Roswell Daily Record. The story was then picked up by a number of newspapers across the country. The story ignited a wave of excitement and speculation about the possibility of extraterrestrial visitors to Earth.

However, within 24 hours, the military leadership, including General Roger Remey, issued a significant retraction, explaining that the debris was merely from a weather balloon, not a flying saucer. The second press release and subsequent retraction fueled intense public fascination and suspicion, laying the groundwork for decades of enduring conspiracy theories.

The Military's Subsequent Weather Balloon Explanation

Following the brief furor, the military released photographs showing officers posing with debris made of foil, sticks, and rubber. These materials were ostensibly the remains of a downed weather balloon. The official narrative later was that the object had been part of a Project Mogul balloon, a then-classified operation designed to detect Soviet nuclear tests by monitoring for sound waves in the upper atmosphere. The secrecy surrounding Project Mogul may have contributed to the confusion and the initial reluctance of authorities to provide detailed explanations. This was what later became the official story.

Personally, I cannot believe that a major in an experienced bombing group could have been fooled by a balloon instead of a flying saucer. As I review the photo taken with Major Marcel holding up the debris, his face appears to me to be holding an obviously fake smile. I believe that he likely felt pretty stupid holding up that balloon material when he knew, for certain, what he had so recently seen with his own eyes and held in his own hands!

For decades, the story of the Roswell incident faded into obscurity, regarded largely as a closed matter. That changed in the late 1970s, when UFO researchers, such as the renowned Stanton Friedman (who was a nuclear physicist), began interviewing people who had been involved. Some former military personnel and even local residents claimed that the material recovered was not of earthly origin. These people described it as exceptionally strong and lightweight, with strange symbols clearly noted. They described what we would know today as a "memory metal" since it could be altered in many fashions yet automatically return to its original configuration. These interviews reignited public interest and spurred a wave of new books and articles questioning the official account.

As more people came forward, stories began to circulate of a massive military cover-up. Some of the witnesses claimed that additional debris and even alien bodies had been recovered and transported to secret government facilities. Witnesses described armed guards, threats of violence, and elaborate efforts to silence those who had seen the wreckage. The narrative shifted from a simple crash to an alleged conspiracy involving the U.S. military, intelligence agencies, and possibly even the president.

The Eventual Cover Story

Finally, nearly fifty years later, responding to persistent public interest and calls for transparency, the U.S. Air Force conducted an "exhaustive review" of the Roswell incident. Their 1994 report concluded that the debris recovered in 1947 was indeed from Project Mogul, a top-secret program involving high-altitude balloons equipped with microphones to detect Soviet nuclear tests. The report emphasized that the materials used in these balloons could easily be mistaken for something extraordinary by those unfamiliar with them.

Another report was later released in 1997. This statement addressed claims that alien bodies had been recovered at Roswell. The Air Force explained that memories of bodies were likely conflated with later incidents during the 1950s, when anthropomorphic test dummies were dropped from aircraft as part of their high-altitude research. These dummies, sometimes mistaken for alien corpses, contributed to the evolution of the Roswell legend. Obviously, the officers in 1947 could not tell the difference between actual beings and artificial figures. Really?

Those must have been some truly life-like dummies. I question exactly where the dummies worked. Were they in New Mexico or in the Pentagon!

The Roswell incident went on to become a staple event of science fiction and UFO literature. There was great interest that rose after the 1980 publication of "The Roswell Incident" by Charles Berlitz and William Moore. It has since been included in many films and TV series, such as "The X-Files" and "Roswell." The story has certainly grown far beyond what the military officers of the original 1947 event would have expected. Countless documentaries, conventions, and tours celebrate the stubborn mystery, drawing visitors to Roswell from around the globe.

The city has seemingly embraced its resulting status as the UFO capital of the world. Each year, Roswell hosts a UFO festival that attracts thousands of enthusiasts from around the planet. The event usually features lectures, costume contests, parades, and reenactments, reinforcing the city's unique place in UFO folklore.

❖ The Barney and Betty Hill UFO Incident

The story of the famous, or infamous, Barney and Betty Hill UFO incident stands out as one of the most influential and widely discussed accounts in the annals of American extraterrestrial lore. Taking place in the early 1960s, it began unexpectedly in the tranquil backdrop of rural New Hampshire. The subsequent tale of the Hills' alleged encounter with an unidentified flying object, and their extraordinary claims later, has sparked extensive public fascination. Since that time, it has continued to inspire prevalent investigations, and helped to shape the modern narrative about the pattern of alien abductions.

Barney and Betty Hill were an unassuming couple living in Portsmouth, New Hampshire. Barney worked for the United States Postal Service, while Betty was a social worker. As an interracial couple (Barney was Black and Betty was White) in the early 1960s, they were already accustomed to navigating societal scrutiny and prejudice. Their story began on the night of September 19, 1961, as they returned from a spontaneous vacation in Niagara Falls and Montreal.

The Night of the Incident

On the night of September 19th, the Hills were driving south on Route 3 through the White Mountains of New Hampshire. Around 10:30 PM, near the town of Lancaster, Betty noticed a strange, bright light in the sky. Initially, the couple assumed it was nothing but a star or a planet. They even speculated that it might be a newly-launched satellite. However, as they drove onward, the light appeared to move erratically, growing larger and closer.

Becoming increasingly curious and uneasy, they stopped the car multiple times to try to get a better look, using binoculars to observe the bizarre object. Betty described it as a craft moving oddly, with multi-colored lights and spinning motions. Barney, initially uncertain, soon grew concerned as the object seemed to follow them on their route, descending and changing direction with uncanny precision.

Eventually, the object reportedly descended directly in front of their car, hovering silently over the highway and filling the expanse of their windshield. Barney exited the vehicle, binoculars in hand, and approached the craft. He later reported seeing humanoid figures through its windows, looking right back at him. Overcome by tremendous fear, Barney fled back to the car, shouting that "They're going to capture us!" The couple sped away down the deserted highway.

Missing Time and Mysterious Symptoms

Upon returning home in the early morning hours of September 20th, the Hills realized that something was amiss. Their journey had taken longer than expected, and both experienced an unexplainable sense of confusion and anxiety. They discovered an odd set of physical symptoms: Betty's dress was torn and stained, Barney's shoes were scuffed, and the strap of his binoculars was broken. Both felt compelled to shower and inspect themselves for unusual marks. Their watches had stopped working, never to run again. Yet they could not understand how any of these signs could have been created. All they had done was to drive home after seeing that craft.

In the days that followed, Betty began having vivid nightmares about being taken aboard a spacecraft and receiving a weird examination by strange beings. Barney, always the more reserved of the two, began to experience bouts of anxiety and suffered unexplained physical discomfort. The couple contacted the Air Force to report what they had seen, and the incident was logged as part of Project Blue Book, the United States' official UFO investigation at the time.

Seeking Answers: Hypnosis and the Birth of Alien Abduction Lore

Haunted by their peculiar experiences and seeking clarity, the Hills were eventually referred to Dr. Benjamin Simon. He was a well-respected psychiatrist and neurologist in Boston. Between January and June of 1964, Dr. Simon conducted a series of regression-type hypnosis sessions with both Barney and Betty. Under hypnosis, the Hills were able to recount remarkably similar stories of being stopped by unknown humanoid beings. They were then led aboard their disc-shaped craft, and were next subjected to what they perceived as disturbing medical examinations.

They described the beings as having large, slanted eyes, grayish skin, and spindly bodies. This description which would later become iconic in accounts of the legendary "Gray" aliens. Betty claimed to have conversed with these beings in English, and even recalled seeing a star map, which she later attempted to draw from memory. Eventually, some amateur astronomers and ufologists would attempt to match Betty's map to known star systems, suggesting (controversially) that it pointed to the star Zeta Reticuli.

Public Reaction and Media Sensation

The Hills' story was first publicly revealed in a series of articles by journalist John G. Fuller in 1965, and later in his book, "The Interrupted Journey," published in 1966 with the Hills' full cooperation. Their unconventional account quickly became a cultural phenomenon, being featured in magazines, television specials, and numerous documentaries. The couple themselves appeared in interviews, speaking candidly about their terrifying ordeal, even as they endured a slew of skepticism and strong ridicule.

The incident was one of the first widely publicized accounts of alien abduction in the United States and helped to crystallize many of the details and motifs. These would include missing time, unusual medical examinations, and telepathic communication with their captors. These elements would soon define the genre in subsequent decades.

Investigations and Skepticism

The use of hypnosis in their case has been particularly controversial, with many critics arguing that it can create false memories, especially when subjects are highly suggestible. Others note inconsistencies between the Hills' conscious recollections and their hypnotic testimony. Other critics point to Betty's revealed earlier exposure to science fiction media as a possible influence on her story.

Nevertheless, some ufologists and supporters argued that the Hills were credible, sincere, and had little to gain from fabricating such a story. This was particularly true when considered with the social racial stigma and the unwanted attention it attracted. To believers, their tale is the very basis of many alien theories.

Legacy and Cultural Impact

The Hills themselves became reluctant celebrities. Despite the skepticism, both maintained the truth of their account until their deaths. Barney died in 1969 and Betty died in 2004. Their reported experiences have been the subject of numerous books, documentaries, and dramatizations, including the 1975 television movie "The UFO Incident," in which their story was brought to a national audience.

More than sixty years after that fateful night in New Hampshire, the Barney and Betty Hill incident continues to captivate the imagination of believers and skeptics alike. It stands as a testament to the enduring power of mystery and the human thirst for answers beyond the stars. Whether the truth lies in the extraordinary, or in the uncharted depths of the mind, it is a fascinating case.

❖ **The Kecksburg UFO Incident: A Comprehensive Exploration**

On a chilly December evening in 1965, the small rural community of Kecksburg, Pennsylvania, was thrust into the limelight of international intrigue and speculation. What occurred on that night has become one of the most enduring and discussed UFO mysteries in American history: the Kecksburg UFO incident. Through eyewitness accounts, official statements, journalistic investigation, and decades of public fascination, this key event has inspired multiple theories, ranging from a simple meteorite fall to the retrieval of alien technology amid Cold War secrecy. Let us delve into the details, controversies, and legacy of the Kecksburg UFO incident.

The Event: December 9, 1965

As dusk settled on December 9, 1965, residents across six states and parts of Canada witnessed a bright fireball streaking through the sky. The object was visible from Michigan to western Pennsylvania, with countless witnesses reporting a glowing, fiery trail. Reports described the object as greenish-orange, moving at a moderate speed, leaving a smoke trail, and emitting sonic booms.

In Kecksburg, a small community about 30 miles southeast of Pittsburgh, residents were startled as the surprising article appeared to come down to the earth in the nearby woods. Some reported a thud or impact, and tales quickly spread of blue smoke rising from the forest. The news travelled quickly throughout the region. Soon thereafter, the rural roads were clogged with curious onlookers, local authorities, volunteer firefighters, and a significant array of law enforcement officers.

Eyewitness Accounts

The first people to reach the scene described a large metallic object, partially embedded in the ground. Most famously, accounts described the object as acorn-shaped, about the size of a Volkswagen Beetle, and covered in strange markings or hieroglyphics around its base. No one had ever seen anything like it!

Some local residents claimed they were stopped by authorities or military personnel before they could approach near to the object, while others insisted they got close enough to see these vital details. In the hours following the crash, the area was reportedly cordoned off by the military, with trucks and personnel arriving swiftly. They shined bright lights, swiftly creating a security perimeter, and asserting total control of the area. Local law enforcement authorities were insistently moved out by the federal authorities.

Differing Descriptions

Witnesses differed on many details, but a common thread soon emerged. This object was unlike any conventional aircraft, or known meteorite track or landing, that even experts would have expected. Multiple early accounts referred to the acorn shape of the craft, having a metallic bronze color, and odd symbols reminiscent of ancient hieroglyphs.

Children playing nearby, volunteer firefighters, and adults alike recounted stories of the intimidation, hurried activity from the military authorities, and the sudden exodus of a large flatbed truck carrying a tarpaulin-covered cargo rushing out of the area late that night.

The Immediate Official Explanation

The U.S. military and NASA quickly responded to the many media inquiries, stating that nothing was recovered and that the most likely explanation was simply a meteor or "bolide" burning up in the atmosphere. Early reports in regional newspapers ran headlines such as "Searchers Fail to Find Object" or "Meteor Startles Six States."

Media Coverage and Public Fascination

Within days, the "Kecksburg Incident" became national news. Radio and TV stations dispatched reporters to the small community. The event's apparent secrecy component, the heavy-handed military presence that was so widely recounted, and the apparent lack of any clear answers only deepened the sense of mystery.

Journalists from outlets like KDKA (Pittsburgh) and the Tribune-Review documented the local residents' stories and attempted to reconcile them with the official statements. Over the years, these early interviews and broadcasts have become foundational material for researchers and serious UFO enthusiasts. The beginning of a legendary event had risen.

Theories and Explanations

Over time, several theories have emerged to explain the Kecksburg UFO incident:

- Meteor or Bolide: The most conventional explanation is that a meteor entered the Earth's atmosphere, producing the fireball seen across a wide area, and disintegrated or impacted in the Kecksburg forest. Some astronomers support this, citing the visible trajectory, sonic booms, and greenish hue.

- Soviet Satellite (Kosmos 96): In the years since, researchers have suggested that the object might have been Kosmos 96, a Soviet Venus probe that failed to reach orbit and re-entered Earth's atmosphere around that time. However, timelines and orbital calculations provided by NORAD and NASA have been inconclusive or contradictory. Along these same lines, a launch from the Baikonur Cosmodrome of an ICBM rocket with a test re-entry vehicle would have closely mimicked the same trajectory this event noted.

- The possibility of an American military experiment or even our creation of a satellite with space debris has been postulated. The Pentagon and NASA were performing all kinds of testing and use of surveillance satellites and similar technology at this time. If the object was truly a piece of classified U.S. military or NASA technology, it would have been of prime importance not to allow our adversaries to know this. The Cold War context, military presence, and extensive secrecy lend some support to this theory, though no official evidence has ever surfaced.

- Extraterrestrial Craft: The most sensational theory posits that the object was indeed an alien spacecraft. After all, a large number of witnesses cited the acorn shape of the article, the unusual markings that were seen, and emphasized the reported recovery operation. These would all work to support this idea. While popular in UFO circles, no definitive evidence supports this claim. But then again, if the evidence had been suppressed, that could be the precise reason no evidence has been reported since the event took place.

- There have been questions about the mysterious Nazi World War II experiment known as "Die Glocke." Die Glocke, which is German for "The Bell," was an alleged top-secret Nazi technology project that has captured the imagination of many folks around the globe. It was described in alternative history and conspiracy literature as an advanced and possibly otherworldly device. The legend of Die Glocke emerged largely from the works of Polish author and journalist Igor Witkowski, who in his 2000 book "Prawda o Wunderwaffe" ("The Truth About the Wonder Weapon"), claimed to have seen classified documents and interviewed former Nazi scientists.

According to Witkowski and later writers, Die Glocke was developed at a facility known as Der Riese ("The Giant") in the Owl Mountains of

Poland. The device itself was supposedly shaped like a large bell, standing roughly 9 to 12 feet tall and about 6 feet in diameter. Encased in a thick metal shell and filled with mysterious swirling fluids termed "Xerum 525," Die Glocke was said to emit potent radiation and possessed properties that defied the conventional laws of physics.

Eyewitness testimonies, which are unfortunately second-hand and of questionable veracity, say that the effects on a person when near the device included bizarre things. These comprised the actual crystallization of plant matter, the gelification of nearby animal tissues, and the severe illness or death among the operational personnel. Skeptics, however, argue that there is no direct evidence for the existence of Die Glocke and that its story is a blend of rumor, misinterpretation, and creative imagination.

While there seem to be superficial similarities, I believe that is where this story ends. The allied armies searched for this particular program, including it in the Operation Paperclip search for advanced Nazi technology. While the jet programs and rocket programs were definitely included in these efforts, there were no reports of finding any confirmed materials related to Die Glocke.

During my time served in West Germany, I came to know many Luftwaffe flight surgeons and engineers that I trusted. They told me about all kinds of planned experiments they had in mind. I was taken to "Fliegerhorst Fürstenfeldbruck" or "Flugplatz Fürstenfeldbruck," as it is known in German. The Fürstenfeldbruck Airbase museum held a variety of strange aircraft that the Nazis had tried to create. Many were very strange, yet I was told that none of them entered service in any meaningful fashion. The problem was that the war had turned against them and they had greatly limited resources available by that time. None of the Luftwaffe officers I met gave any credence to the successful use of that bizarre Die Glocke program. Therefore, I am also reluctant to give any credibility to the experimental program Die Glocke in reference to the Kecksburg incident.

Government Secrecy and Legal Battles

A major source of public suspicion has been the perceived lack of transparency from government agencies. This is not unique to this event. In the early 2000s, investigative journalist Leslie Kean, aided by the Sci-Fi Channel, filed Freedom of Information Act (FOIA) lawsuits against NASA for any records related to the Kecksburg incident. After years of legal wrangling, a judge eventually ordered NASA to release documentation.

However, the information provided by the space agency was rather scant and entirely inconclusive. NASA's official statement eventually claimed that their experts had fully examined the site and recovered no space debris. This statement would not, in any case, indicate whether the military had recovered any objects or debris. The NASA account, therefore, only fueled more speculation and encouraged the allegations of a government cover-up. The U.S. Air Force maintained its original stance that nothing of interest was found at the site.

Legacy and Cultural Impact

Kecksburg has embraced its place in UFO lore. In a similar pattern to Roswell, this community hosts an annual UFO Festival, complete with lectures, parades, and even a replica of the mysterious acorn-shaped object. The event has become an integral part of the town's identity, drawing tourists, researchers, and believers from around the world.

The incident has been widely featured in numerous documentaries, TV programs (including "Unsolved Mysteries"), a host of books, and been reported in many UFO podcasts. Its enduring allure lies in the combination of eyewitness testimony, incredible government secrecy, and the persistent absence of definitive proof.

Media Perspective

Go back to the 1938 "War of the Worlds" broadcast by Orsen Welles. While the program was introduced as merely a dramatic adaptation, the format and timing of the news bulletins led many listeners to believe the terrifying events being described were real. Some listeners had missed the opening disclaimer while others tuned in after the broadcast had begun. For these individuals, the succession of realistic news flashes, reporting Martian heat rays, poisonous gas, and a military mobilization, sounded very authentic and quite terrifying.

The exact extent of the panic induced by that unique broadcast has been debated by historians. Contemporary accounts included reports of people fleeing their homes, clogging highways, and flooding police stations with frantic calls. Newspapers and rival radio networks quickly capitalized on the story, publishing sensational headlines about mass hysteria.

Nevertheless, the incident remains a tantalizing piece of American folklore and a focal point for debates about government secrecy, the possibility of extraterrestrial visitation, and the power of collective mythmaking.

The Kecksburg UFO incident still stands at the intersection of unsubstantiated fact, wild folklore, and as an enduring mystery. Whatever truly happened in the Pennsylvania woods in 1965, the event continues to captivate imaginations, fuel debates, and inspire curious inquiries into the unknown. Whether a falling meteor, a lost satellite, or something altogether stranger, the Kecksburg incident remains a testament to humanity's fascination with the unexplained and the enduring search for truth beyond the horizon.

❖ The Malmstrom UFO Missile Incident

On March 16, 1967, reports emerged from personnel stationed at Malmstrom Air Force Base regarding sightings of unidentified flying objects near their installation. These objects were described as glowing orbs, hovering silently and performing movements that defied conventional aviation capabilities. Witnesses included security guards and other military personnel, individuals highly trained to identify aerial phenomena and threats.

The inexplicable appearance of these objects coincided with a series of malfunctions in the missile systems at the Oscar Flight Launch Facility, part of Malmstrom's sprawling network. According to firsthand accounts, as the UFOs remained in the vicinity, ten different thermonuclear missiles went offline, rendering them completely inoperative. This was not a simple technical glitch. Later investigations confirmed that the shutdown had no identifiable cause and was not the result of human error or environmental factors. It was upsetting to have a single missile go offline. To have ten of the missiles to do this at nearly the same time was completely terrifying!

Safety Factors Used to Safeguard the Minuteman ICBMs

It was at the highest levels of security that we protected these missiles and ensured operational readiness. Let us think about why. At the time, the Minuteman intercontinental ballistic missiles (ICBMs) represented a cornerstone of the United States' strategic nuclear defense systems. Along with bombers that could carry nuclear bombs and the missile-launching submarines hiding in the world's oceans, these ICBMs were the third leg of the "Nuclear Triad."

There were intense physical security measures for all of our strategic forces, and these missiles were certainly no exceptions. The physical safeguarding of Minuteman ICBMs was paramount to the deterrence factor. The missile silos, which housed the ICBMs, were distributed across secure and heavily fortified locations. These silos were designed to withstand significant external threats, including conventional weapon attacks and natural disasters. To prevent unauthorized access, the silos and their deeply buried control centers were surrounded by multiple layers of security, such as:

- Perimeter Fencing: High-security fencing equipped with intrusion detection systems ensured any unauthorized entry attempts would be immediately detected.

- Surveillance Systems: Advanced surveillance technologies, including cameras and motion sensors, continuously monitored the missile sites.

- Armed Security Personnel: Specially trained military guards were stationed at and near the sites to respond swiftly to any potential threats.

Command and Control Safeguards

The command-and-control systems of the Minuteman ICBMs were designed with redundancy and security to prevent unauthorized launches or communication breaches.

- Two-Person Rule: A strict protocol ensured that no single individual had the authority to initiate a launch. This rule mandated that two authorized personnel must verify and execute launch procedures simultaneously.

- Authentication Protocols: Complex codes and authentication processes were required to verify the validity of launch orders and guard against false commands.

- Secure Communication Channels: All command communications were encrypted and transmitted via highly secure channels to prevent interception or tampering.

Technological Safeguards

In addition to physical and procedural security, technological measures played a significant role in safeguarding Minuteman ICBMs.

- Sealed Missile Silos: The missiles were housed in sealed silos, which could only be accessed through authorized mechanisms.
- Environmental Controls: Advanced environmental control systems prevented overheating or degradation of missile components, ensuring their long-term operational readiness.
- Fail-Safe Mechanisms: Multiple fail-safe systems were in place to abort a launch sequence in case of errors or unauthorized commands.

Cybersecurity Measures

With the increasing reliance on digital systems, cybersecurity was an essential aspect of safeguarding the Minuteman ICBMs. Rigorous measures were implemented to protect them against any type of cyber threats:

- Firewalls and Intrusion Detection: Cutting-edge cybersecurity tools prevented unauthorized access to the command-and-control systems.
- System Isolation: Critical systems were isolated or air-gapped from external networks to minimize the risk of hacking.
- Regular Security Audits: Frequent audits and updates were conducted to address emerging cyber vulnerabilities.

Training and Human Factors

Human operators played a vital role in the safeguarding process. Comprehensive training programs ensured that personnel were well-prepared to manage any situation involving Minuteman ICBMs. For this reason, safety and professionalism were paramount.

- Regular Drills: Simulated scenarios helped personnel practice their wartime responses to potential threats or strategic emergencies.

- Psychological Screening: Personnel with access to these critical systems were required to undergo periodic psychological evaluations to ensure their reliability and mental stability. A nuclear missile silo was no place for anyone with a disturbed psyche.

- Continuous Education: Ongoing training ensured that all personnel were up to date with the latest protocols and technologies. These were being developed and updated on a constant basis due to the responsibilities these Air Force personnel held within their official duties.

Conclusion

Malmstrom Air Force Base was certainly a key component of the U.S. military's strategic deterrence during the Cold War. Housing Minuteman intercontinental ballistic missiles (ICBMs), this facility represented a critical node in America's ability to respond to threats with nuclear force. Each missile was maintained under stringent security measures, ensuring its readiness for deployment at all times. The idea that such systems could be compromised by external forces, whether they were human or otherwise, was an alarming concept to military planning officials and policymakers alike.

One of the most notable figures associated with the Malmstrom incident is retired Air Force Captain Robert Salas. Salas, who was on duty at the time, has spoken publicly and extensively about the event. He recalls being informed by security personnel of strange objects hovering above the base and moments later receiving reports of missile systems becoming disabled. Salas has insisted that the incident was not only real but involved forces beyond any known terrestrial technology.

Since I have personally met with and held discussions with Mr. Salas, I completely believe his story.

Moreover, the incident raises questions in my mind about the vulnerability of the many nuclear installations worldwide. If UFOs were indeed responsible for the system failures at Malmstrom, what prevented similar occurrences to happen at other bases, or even in other countries? The possibility that such phenomena could bypass human defenses and safeguards remains a very troubling prospect.

❖ The USS Nimitz Encounters of 2004

An Exploration of Unexplained Aerial Phenomena and Naval Encounters

In recent years, UFOs/UAPs near military installations, particularly aircraft carriers, has captured widespread attention. Reports of UAP sightings by U.S. Navy personnel have ignited debates among scientists, military experts, and the public alike. Advanced radar technology, often stationed aboard nuclear aircraft carriers or held within support vessels, has played a pivotal role in detecting and recording these unexplained aerial phenomena. These noteworthy encounters, with their surreal and enigmatic details, severely challenge our conventional understanding of aviation and physics.

The Critical Role of U.S. Navy Nuclear Aircraft Carriers in Modern Global Defense

Aircraft carriers, and their support vessels, are among the most sophisticated and strategically vital assets in modern naval operations. What makes them exceptionally valuable is their ability to relocate to troubled spots wherever they might arise anywhere in the world. Modern radar support vessels are marvels of naval engineering, often serving as multi-role defense platforms. These vessels combine powerful radar sensors with missile defense, command and control, and electronic warfare capabilities. Our carrier groups must be fully protected!

Key Features Include:

- Advanced Phased Array Radar: Modern ships often feature active electronically scanned array (AESA) radars capable of tracking hundreds of targets simultaneously over vast distances.
- Integrated Missile Defense: Vessels like the U.S. Navy's Aegis cruisers and destroyers can detect, track, and intercept ballistic and cruise missiles, forming an extensive protective shield around the carriers they defend.

- Networked Operations: Through secure data links, radar support vessels share their key targeting information with other ships, aircraft, and command centers, forming a real-time operational picture.

- Stealth and Survivability: Modern ships incorporate stealth design to reduce detectability and employ effective countermeasures to survive within very hostile environments.

Examples of Radar Support and Ballistic Missile Defense Ships:

- Ticonderoga-class cruisers and Arleigh Burke-class destroyers of the US Navy: Equipped with SPY-1D radars and the Standard-class missile interceptors, these vessels serve as the backbone of U.S. carrier strike group air defense. Enabling considerable missile defense systems to be available to protect our carriers, these are vital assets, regardless of where in the world we might have to deploy these carrier battle groups.

- Type 45 Destroyers (UK): These are the analogous British Royal Navy's premier air-defense destroyers and present massive upgrades to earlier systems. The destroyers feature the SAMPSON radar system and Sea Viper missile system. Both are considered world-class.

The Tactical Significance of Radar Vessels

Radar picket ships enhance the defensive perimeter around an aircraft carrier, allowing for:

- Early engagement of incoming threats, which maximizes the reaction time for the carrier's air wing and defensive weapons.

- Layered defenses, with overlapping fields of radar coverage and increased overall missile interception capabilities.

- Coordination of friendly fighters and surface-to-air missile batteries for optimal resource allocation.

- Decoy and deception roles, sometimes drawing enemy fire or confusing adversaries with false electronic signatures.

There have been a number of incidents when these important support vessels have encountered the same UAPs as their carriers and flight crews. Because of this, data collected from aircraft carriers and their support vessels utilize the latest radar systems that are unparalleled in their precision. These radars can detect objects traveling at extraordinary speeds or maneuvering in ways that defy known aeronautical capabilities. It is within this context that any UAP sightings near such vessels acquire a layer of credibility that might otherwise be absent.

Historical Context

All the way back to October 11, 1492, sailors have witnessed strange lights and objects they could not explain. In this case, it was Christopher Columbus who was sailing west, seeking a new trade route to Asia. Encountering these unexpected sights, Columbus documented these UFOs in his official ship's log. He would not be the last sailor to do this.

During 2004, the USS Nimitz Carrier Strike Group was operating off the coast of California, performing combat exercises. From out of nowhere, a radar return detected an unidentifiable object in their airspace. A pair of

F/A-18 Super Hornet fighter jets from the Nimitz were dispatched to intercept one of the unidentified objects. Piloting one pair of the jets were Commander David Fravor and Lieutenant Commander Alex Dietrich. What they encountered remains one of the most extraordinary narratives to emerge from the event.

Fravor described seeing a white, oblong-shaped object, approximately 40 feet in length, hovering above the ocean. The object moved in ways that defied the laws of physics as we understand them. It would accelerate instantaneously, change direction without losing momentum, and seemed to operate without any visible means of propulsion. Fravor compared its movements to a "ping-pong ball bouncing inside a glass."

Dietrich recounted the incident as well. In describing its flight characteristics, she stated, "No predictable movement, no predictable trajectory."

This UAP showed amazing abrupt changes in direction, hypersonic speed, and had the complete absence of any visible propulsion systems. Radar operators aboard the USS Princeton, a guided-missile cruiser accompanying the carrier, fully corroborated these observations with precise radar data that confirmed the object's anomalous movements.

As Dietrich circled above, Fravor went down for a closer look. He said the object was about the size of his F/A-18F, with no markings, no wings, and no exhaust plumes. As soon as Fravor tried to cut off the UAP, it accelerated so quickly that it seemed to disappear, he recalled. Seconds later, the USS Princeton reacquired the UAP on its radar. It was approximately 60 miles away.

As Fravor attempted to approach the object, it seemed to react to his presence, darting away at incredible speed. When the aviators returned to the carrier, they recounted their experiences, which were corroborated by radar data and infrared camera footage captured by another aircraft.

Video Evidence

Further evidence came in the form of a video recorded by one of the fighter jets' infrared camera systems. Known as the "FLIR1" footage, this video shows an object moving at high speed while pilots candidly discussed its unusual behavior. The video was later declassified and released to the public, igniting widespread media coverage and speculation.

Scientific and Military Reactions

Both scientists and military personnel have offered varying interpretations of the encounters. Some suggest that the objects were advanced experimental aircraft, possibly from foreign nations, or even from American contractor companies, while others hypothesize that they were drones or natural phenomena misinterpreted by radar systems.

However, the lack of evidence supporting terrestrial explanations has led many to speculate about non-human technologies. The incident has been cited in congressional hearings on UAPs and has contributed to the Pentagon's decision to establish the Unidentified Aerial Phenomena Task Force. Aviators from the carrier were dispatched to investigate and reported seeing a white, oblong craft resembling a Tic Tac mint. Once identified by flight crews, the object exhibited superior flight behaviors that were inconsistent with any known aircraft. And it was not alone. Radar recorded many other UAPs in their same vicinity.

What about reports that this UAP was actually an American contractor-related craft?

There is no publicly available evidence to confirm that Lockheed, or its aerospace division Lockheed Martin, was responsible for the creation of the so-called Tic Tac craft. Despite recent speculation surrounding major defense contractors, especially Lockheed Martin, the true origin or manufacturer of the craft remains officially unknown.

This would not be unprecedented for this defense contractor. They were, after all, the developers of the S-3 Viking. This is Lockheed's carrier-based workhorse aircraft. Developed in partnership with Vought, the S-3 Viking first flew in 1972 and entered service with the U.S. Navy in 1974 as an anti-submarine warfare (ASW) aircraft. The S-3 Viking was specifically designed for carrier operations, featuring folding wings, a tailhook, and robust landing gear for catapult launches and arrested landings.

The S-3 Viking proved to be an important and versatile platform:

- Anti-Submarine Warfare: Equipped with sonobuoys, magnetic anomaly detectors, and torpedoes, the S-3 was the backbone of carrier-based ASW for decades.
- Surface Surveillance: Advanced radar and sensor systems allowed the aircraft to perform maritime patrol and surface search missions.

- Aerial Refueling: Many later S-3s were converted to "buddy tankers," extending the range of other carrier-based aircraft.

- Electronic Warfare: Some variants provided advanced high-tech electronic surveillance, jamming capabilities, and aerial reconnaissance.

The S-3 Viking remained in frontline service until 2009 and continues to serve in secondary roles, such as testbed platforms, to this day.

Advanced Projects and Technology Demonstrators

Through the late 20th and early 21st centuries, Lockheed Martin has engaged in numerous advanced projects, such as the U-2 and the SR-71. They have also created vital aircraft to support naval aircraft carrier operations:

- Stealth Aircraft: Lockheed Martin's expertise in stealth technology, exemplified by the F-117 Nighthawk and the F-22 Raptor, influenced naval aviation. Although these specific models were not carrier-based, their design and technology paved the way for future naval adaptations.

- F-35C Lightning II: As the lead contractor of the F-35 program, Lockheed Martin developed the F-35C variant specifically for carrier operations. The F-35C features larger wings, reinforced landing gear, a tailhook, and avionics tailored for the challenges of launch and recovery at sea. It represents the most advanced multirole stealth fighter ever fielded on an aircraft carrier.

- Unmanned Aerial Systems (UASs): Lockheed Martin has invested in the development of unmanned systems capable of carrier operations. Projects like the Sea Ghost, a stealthy carrier-based UAS proposal, demonstrate Lockheed's commitment to next-generation naval aviation.

Some theories have suggested that advanced aerospace contractors, such as Lockheed Martin, might be involved in continued secret government projects that might be associated with UAPs. Many people may not realize this, but the Air Force builds <u>no</u> aircraft. They have purchased each aircraft they own from a defense contractor. It is the contractors, such as Lockheed Martin, that have the expertise to build these advance fighters, bombers, tankers, etc. However, no official statements or credible information have been released that directly links Lockheed Martin to the Tic Tac craft. This has not at all stopped the recurring speculation that this organization is the actual source for this craft.

Another Key Incident

One of the other reported incidents involving Navy radar and UAPs occurred in 2015 during training exercises off the East Coast of the United States. Fighter pilots from the USS Theodore Roosevelt reported multiple encounters on multiple days with unidentified objects that appeared on their radars and were also caught on video by their infrared cameras. Known as the "Gimbal" and "GoFast" videos, these documented encounters showed objects with no visible means of propulsion that were moving at extraordinary speeds.

The naval aviators described these objects as flying at altitudes ranging from sea level to 30,000 feet and performing maneuvers that defied current aerodynamics. Their radar systems, including the AN/SPY-1, a highly advanced radar used in Aegis Combat Systems, confirmed the presence of these objects. The data was later analyzed by the Pentagon's Advanced Aerospace Threat Identification Program (AATIP), adding to the growing body of evidence surrounding UAPs.

The Implications of UAP Encounters and Our Nuclear Aircraft Carriers

The repeated detection of UAPs near U.S. Navy aircraft carriers raises several questions of strategic importance. First, there is the issue of national security. Are these objects some form of foreign surveillance devices, adversaries' advanced drones, or something else entirely? The U.S. Navy has taken these sightings seriously, implementing new protocols for their aviators and crews to report such encounters without the previous fear of stigma.

Second, there is the scientific curiosity surrounding these phenomena. If the objects detected are not of Earthly origin, they could represent a breakthrough in understanding propulsion, energy, and material technologies. Such discoveries could revolutionize multiple scientific fields, from aerospace engineering to experimental physics.

Lastly, these incidents have a cultural and philosophical impact. The possibility of extraterrestrial involvement in these encounters challenges humanity's understanding of our place in the universe. The Navy's radar data serves as a critical tool in exploring these profound questions, offering a glimpse into the mysteries of the skies.

Conclusion

The phenomenon of UAPs near our Navy's nuclear aircraft carriers represents a fascinating intersection of advanced technology, unexplained phenomena, and strategic importance. While much remains unknown to us, what is clear is that the Navy's radar systems are playing an indispensable role in further documenting these occurrences.

As more data is collected and analyzed, the hope is that humanity will come closer to understanding the true nature of these mysterious objects. Whether they turn out to be our own unconventional craft, a foreign adversary's advanced technology, a natural atmospheric anomaly, or evidence of extraterrestrial life, their persistent presence near aircraft carriers demands continued scrutiny and investigation.

The Department of Defense needs to know whether these anomalous craft form any strategic threats to our national defense!

Legislative Developments

By 2020, growing interest in UAP phenomena led to the inclusion of UAP-related mandates in defense bills. The U.S. Congress instructed the Department of Defense to provide detailed reports on their UAP encounters, urging increased transparency and further study.

❖ **Unidentified Submerged Objects, or USOs, and Their Connection to UAPs**

Exploring the Mysteries Below the Surface

Unidentified Submerged Objects have been receiving increased analysis over the last decade or two. USOs are enigmatic phenomena akin to UFOs, except that their domain is underwater as opposed to the skies. While UAPs have captivated human imagination for decades, USOs remain a shadowy counterpart, tethered to the deep. The study of USOs has slowly gained more traction among researchers and enthusiasts. Often viewed as connected to the broader field of unidentified aerial phenomena, USOs are being increasingly scrutinized.

What Are USOs?

USOs are objects or phenomena observed beneath water surfaces that defy any conventional explanation. Like UFOs, they may exhibit advanced mobility, high speeds, and erratic movements that challenge the capabilities of known human-made technologies. Reports of USOs often include descriptions of glowing shapes, rapid ascents to the surface, or seamless transitions between water and air. In many cases, USOs become UFOs, and vice versa.

Although less frequently documented than UFOs, USOs have been reported in a large number of oceans, seas, lakes, and even rivers worldwide. Their elusive nature raises intriguing questions about their origin, purpose, and potential connection to extraterrestrial entities.

Historical Accounts of USOs

The phenomenon of USOs is not new. Ancient maritime lore includes tales of glowing or fast-moving objects beneath the waves, often interpreted as mystical or divine at the time. However, modern accounts provide more detailed descriptions and are often corroborated by radar, sonar, or eyewitness testimonies.

Military Encounters

Military personnel have reported several encounters with USOs, particularly in naval operations. For instance, during the Cold War, both American and Soviet forces documented instances of mysterious underwater objects that outpaced their submarines or evaded sonar detection. These sightings often occurred in strategic areas such as the Arctic Ocean, where the depths of the seas are less explored.

Civilian Reports

Civilians, including sailors and deep-sea divers, have also reported USO sightings. Many of these accounts share common themes, such as sudden appearances, luminous undersea glows, or objects transitioning between water and air without turbulence or resistance.

USOs and UFOs: The Connection

USOs and UFOs are often discussed within the same framework due to their shared characteristics, such as advanced propulsion systems, rapid movements, and unexplained origins. Some theories suggest that USOs might be underwater bases or crafts operated by the same entities responsible for UFOs. These craft might use advanced technologies to navigate seamlessly between air and liquid mediums, as suggested by certain eyewitness accounts of objects plunging into the ocean without even creating a splash. Many researchers believe that USOs and UFOs are actually one in the same.

The Hypothesis of Underwater Extraterrestrial Bases

One of the more popular theories posits that extraterrestrials may use underwater bases as a means of avoiding human detection. The vast, uncharted expanses of Earth's oceans provide ideal locations for such facilities, shielded from probing satellites and terrestrial interference. With the expanse of the oceans across the planet, there could be many places for them to hide.

Scientific Perspectives on USOs

Scientific explanations for USOs range from misinterpretations of natural phenomena to advanced military technologies. For example, bioluminescent marine organisms or underwater seismic activity might create illusions mistaken for USOs. On the other hand, experimental submarines or drones might account for some sightings.

Limitations in Research

Research into USOs is hindered by a lack of funding, public interest, and technological capabilities. The ocean remains one of Earth's least explored terrains, with more than 80% of it still unmapped. This makes the investigation of underwater phenomena particularly challenging compared to aerial phenomena.

USOs in Popular Culture

USOs have also made their mark in popular culture, appearing in books, movies, and documentaries. Films like "The Abyss" (1989) and recurring themes in science fiction literature often explore the idea of advanced underwater civilizations or alien technologies hidden beneath the waves.

These artistic expressions help keep the mystery alive, fueling public curiosity and debate.

Modern-Day Investigations

With the advent of better technology, including high-definition underwater cameras and advanced sonar systems, investigations into USOs are gradually improving. Organizations studying UFOs, such as the United States Pentagon's UAP Task Force, have expanded their remit to include USOs under the broader category of UAPs.

Although USOs remain one of the lesser-known mysteries of our planet, their study is increasingly recognized as vital in understanding unexplained phenomena. Whether they are natural occurrences, secret human technologies, or evidence of extraterrestrial life, USOs challenge our understanding of the world beneath the waves. As research progresses, they may ultimately help us unlock the secrets of the deep and perhaps even assist us in answering the enduring question of humanity's place in the universe.

❖ **Another of My Stories:**

In my role as a flight surgeon at the 45th Space Wing, I also supported the DDMS mission of DoD operations worldwide. This was to directly promote the requirements needed to assure the safety capabilities needed for the NASA Space Transportation System, or STS (otherwise known as the space shuttle program).

Understanding Military Contingency Planning and Operations for the Space Shuttle Program

The Space Shuttle operational era, from 1981 to 2011, demanded one of the most sophisticated and far-reaching contingency planning systems in the history of human spaceflight. At the heart of this effort was the previously mentioned Department of Defense Manned Space Flight Support Office. DDMS was tasked with developing and coordinating an array of contingency modes and support mechanisms to ensure the safety of astronauts and the retrieval of the orbiter and crew, no matter where or when an emergency could occur.

Introduction to the DDMS

The Department of Defense Manned Space Flight Support Office was originally established in the late 1950s to provide support for NASA's planned human spaceflight programs. DDMS grew to prominence especially during the Apollo and Space Shuttle programs. The offices of the DDMS worked in close coordination with NASA's Mission Control Center at Johnson Space Center in Texas, the Kennedy Space Center in Florida, rescue organizations globally, all military branches, and in coordination with many international partners. Its primary objective during the time of my service was to prepare for and coordinate rescue and recovery operations in the event of a Space Shuttle emergency. Unfortunately, shuttle emergency landings could possibly involve both the United States and potential landings anywhere else around the globe.

Space Shuttle Contingency Modes: An Overview

Space Shuttle launches and missions faced a spectrum of potential emergencies, ranging from launch pad aborts to in-flight anomalies and off-nominal landings. As such, a family of contingency modes was clearly defined, with each one having specific procedures, hardware requirements, and support teams. Broadly, these could be categorized as:

- Return to Launch Site (RTLS): If an early shuttle mission abort was to occur in the first few minutes of flight, the orbiter would need to separate from the two solid rocket boosters (SRBs), which were the two tall white solid rockets attached to each side of the external tank. In two minutes and five seconds after launch, the SRBs would separate and allow for the potential RTLS. The large orange External Tank was filled with liquid hydrogen and liquid oxygen. These super-cold liquids were what fueled the orbiter's three space shuttle main engines (SSMEs). The continuing thrust provided by the SSMEs would allow the RTLS crew to retrace their launch path back toward a landing at Kennedy Space Center.

Once the crew was able to enter a safe corridor for actions, they would then discard the external tank. From this point, they would hopefully glide back to a safe landing at the Shuttle Landing Facility (SLF). My helicopter crews and I would be waiting at the SLF, just in case any problems might arise. After any failed launch attempt with the crew still in the orbiter, we would not be released from the SLF until the astronauts had been safely extracted from the spacecraft and returned to their crew quarters.

- Transoceanic Abort Landing (TAL): If a space shuttle and crew had flown farther out over the Atlantic Ocean after launch, they would be too far away from any return to KSC. Still, if they were not high enough or fast enough, they would be unable to reach a safe orbit. In this case, their only option would be to land on the eastern side of the Atlantic. During my time, we maintained four TAL sites. These were the Morón and Zaragoza Air Bases in Spain, the Ben Guerir Air Base in Morrocco, and the Yundum International Airport in The Gambia.

- Abort Once Around (AOA): If a shuttle crew was able to gain enough speed and altitude to reach an orbit, but did not feel it was safe to continue their mission, they could use this abort mode. They would fly far enough during their single orbit to reach a location to perform a safe de-orbit burn. This likely would be selected to aim for a landing at Edwards Air Force Base (what NASA preferred to designate as the Dryden Flight Research Center) as the primary return site. Second, the crew might be able to safely return to the SLF at Florida's Kennedy Space Center. The tertiary landing site would have been the White Sands Space Harbor in New Mexico. As an action of last resort, there were emergency landing sites identified at various airports around the world.

- Abort to Orbit (ATO): An abort to orbit was available when a space shuttle crew was high enough to reach a stable orbit, even if it would be lower than the intended orbit planned by NASA. This ATO contingency was actually used on July 29, 1985. The orbiter

Challenger had its center SSME malfunctioning and it had to be shut down only five minutes and 46 seconds after liftoff. The lower orbit it was able to achieve was still satisfactory for the mission and the crew had a safe landing eight days later on August 6, 1985.

What made this mission especially interesting to me was that when I arrived at Patrick AFB, the commander was Colonel Roy D. Bridges Jr. He was a former astronaut who was on that very same mission, STS-51-F. He happened to live in military housing directly across the street from me. He was a very fascinating man to speak with. He later became a general who finished his remarkable career at the Air Force Materiel Command, located at Wright-Patterson Air Force Base, Ohio.

- Contingency Landing at Emergency Landing Sites: As I mentioned above, this was a choice of final resort. There were a large number of airports having a runway long enough to land the space shuttle, in the gravest of emergencies. No one wanted to require this option for several reasons. Among them would be the hazards of the chemicals emitted by a landed orbiter. There would also be the difficult logistics of creating suitable capabilities to load that orbiter onto the back of the 747 Shuttle Carrying Aircraft. Any airport receiving the shuttle would have its flight routine greatly affected by this landing for days, if not weeks.

- Search and Rescue (SAR) contingencies

For adverse land crashes, water landings, or astronaut ocean bail-outs unavailable at the routinely prepared sites, the DDMS would have immediately coordinated with any forces available involving the U.S. Coast Guard, Army, Navy, Air Force, and even included international SAR agencies. Assets at the primary, secondary, or tertiary launch or landing sites included helicopters, fixed-wing aircraft, ships, and pararescue teams/flight surgeons. The medical assets were all trained and certified in astronaut extraction, emergency medical stabilization, and proper transport decisions for selecting the appropriate hospitals for definitive medical care.

It was for this latter reason that I deployed to The Gambia for reviewing the medical capabilities at this TAL site.

It was probably the third or fourth day I was in The Gambia and I had converted some dollars to their local currency, which was known as Dalasi. I wanted to buy eight post cards to send to relatives in the States, just for fun. Well, I only had enough Dalasi for the post cards, but the man selling them adamantly insisted that I had to purchase eight post card stamps as well. Not having enough Dalasi to do so, I assured him I would simply buy these at the hotel in which I was staying. He became very disturbed, but I left the shop anyway.

Upon arriving to the hotel, I tried to buy stamps at the front desk for my post cards. The man at the desk then informed me it was illegal for him to sell me eight stamps, unless I was also willing to purchase eight more post cards. No matter what I said, the desk clerk would not relent. I then went to the hotel next door, where I received the same treatment. I next walked across the paved road onto a dirt road where there was a supermarket, as they called it. It was really the size of a small American convenience store. A man with a pushcart in the middle of the road asked if I wanted to buy cigarettes, including American ones. I loudly told him all I wanted to do was to purchase eight stamps for post cards.

That was when a young boy approached me and explained that his uncle could sell me stamps. I reiterated that I only wanted stamps, since I already had bought the post cards. The boy said this was no problem, so I immediately followed him across another dirt road into a local shopping square, perhaps forty yards by thirty yards. The eager guide took me completely through the main area to a small gate leading out from this area.

My safety antennae went up because I saw no buildings beyond that gate, so I halted right there. Just then, the boy brought his "uncle" to me, and I saw that he did not look like a normal shopkeeper. The man spoke to me in broken English, saying, "You no want stamps. You want cocaine!"

I instantly refused this, replying, "I had wanted to buy eight stamps, but I have changed my mind."

The uncle leaned closer to my face and stated, "We have the best cocaine. It comes from Columbia!"

I answered, "I don't care where it is from. I do not want any cocaine."

Things got much worse when I found two men brandishing machetes that had come behind me to the left and to the right.

The uncle declared, "It is the best cocaine you can buy. It is better than any you have ever had before."

Searching behind me, I could see that other people from the shopping area had closed in behind me and that probably no one from the road could even see I was there in the midst of this crowd. I shook my head and said, "I have never tried cocaine, and I do not wish to try it now!"

The uncle smiled, despite a few missing teeth. He promised, "Try it and you will like it. You will <u>like</u> it!"

I had concluded that the longer I stood there the worse my situation was becoming. I looked at the uncle then back to the two machete men. I loudly proclaimed, "I am an American Air Force officer, and I am in-country to support NASA operations here."

I wanted to raise the stakes for these men while they had to consider whether they wanted to harass an American military officer. Pestering a tourist is one thing. Endangering an officer of the United States Air Force was an entirely different challenge. At least, I hoped it was!

I placed my hands between the two machete guys and boldly pushed them apart. As I strode forward with absolute purpose, I forcefully specified, "I have additional American military men over at that hotel. I am going there right now and do not want to be hindered!"

Secretly, I was praying, "Dear God, do not let them have firearms!"

The cocaine men looked at each other as though uncertain how to proceed. Not wishing to allow them any further time to decide on their actions, I swiftly walked toward the hotel.

The uncle then ran to me and argued, "I am your friend. I do you favor. You do nothing for me. What favor you do for me?"

I pulled a five Dalasi bill from my pocket and handed it to him. That was the equivalent of less than half of a dollar in American money. I told him, "Take this, and it is the only favor I will do for you. I will do nothing else!"

Thankfully, he grabbed the five Dalasi bill and smiled. "You my friend. Come back and I will give you big deal on cocaine!"

I ignored this and turned back toward the hotel. I was moving as fast as I could without running. I did not look back until I got on the grounds of the hotel. I was finished shopping for the moment!

The next day, I went to what I felt was a safe shop in Banjul and bought eight post cards, along with eight stamps. I had assuredly learned my lesson!

Chapter 2

What did Ancient Cultures Believe?

❖ **UFOs in the Bible: Ancient Mysteries and Modern Interpretations**

Throughout history, the Bible has been a source of both spiritual inspiration and curiosity regarding the mysteries of the universe. Among the enigmatic topics that have captured the imagination of modern readers are its accounts of unusual celestial events and objects that some interpret as references to UFOs. While the term "UFO" is decidedly modern, many have wondered whether ancient texts, particularly the Bible, contain descriptions of phenomena that could suggest encounters with visitors or technologies that are beyond human comprehension.

Let us explore passages within the Bible that have been interpreted by some as potential references to UFOs. We will analyze their context, and examine both traditional and alternative theories about their meaning. The intention is not to assert a definitive answer, but rather to illuminate the intersection of faith, history, and the ongoing human quest for understanding the unknown.

What Were Possible UFOs in the Minds of the Biblical Writers?

In the context of biblical times, ancient peoples might have described the appearance of strange lights, moving objects in the sky, or celestial manifestations in the language and concepts familiar to them. These would often be framed in their religious, metaphorical, or mythological terms. They simply had no words available to them in order to describe the current elements of our own modern technology. They had no understanding for references to aircraft, rockets, spacecraft, lasers, etc.

Key Biblical Passages Associated with UFO Interpretations

The Vision of Ezekiel (Ezekiel 1)

Perhaps the most cited passage regarding possible UFOs is found in the Book of Ezekiel. In the first chapter, Ezekiel describes a dramatic vision by the Kebar River:

- He sees a "windstorm coming out of the north, with an immense cloud with flashing lightning and surrounded by brilliant light." (Ezekiel 1:4, NIV)

- Within the fire, Ezekiel describes "four living creatures," each with four faces and four wings.

- He then sees "a wheel intersecting a wheel," and these wheels are "sparkling like chrysolite."

- The wheels are described as being able to move in any direction without turning as they went, and they "sparkled like topaz."

Some contemporary theorists, such as Erich von Däniken, have argued that Ezekiel's vision could be interpreted as an ancient encounter with advanced technology. Perhaps this was a spacecraft, with the "living creatures" being its occupants or operators. Supporters of the UFO hypothesis point to the detailed, mechanical-sounding descriptions of the "wheels within wheels" and the way they were described to move.

However, traditional religious interpretations view this vision as a symbolic representation of the glory of God, employing vivid imagery to convey a supernatural experience far beyond ordinary human perception. Nevertheless, if we saw this same vision today, would we be able to use any of our technological terms to identify it more clearly?

The Chariot of Fire (2 Kings 2:11)

Another intriguing account appears in 2 Kings, where the prophet Elijah is taken up into heaven:

- "As they were walking along and talking together, suddenly a chariot of fire and horses of fire appeared and separated the two of them, and Elijah went up to heaven in a whirlwind." (2 Kings 2:11, NIV)

For ancient readers, the "chariot of fire" was a powerful symbol of divine intervention. Some modern UFO enthusiasts, however, speculate whether the account might suggest an encounter with a craft of otherworldly origin, due to the emphasis on fire, motion, and ascension into the sky. When I have watched and listened to space shuttle launches, I found myself wondering how people from those ancient times would possibly describe the liftoff of a modern spacecraft. Seven people in strange orange clothing climb into this massive creation, it roars to life, then carries that crew to space on amazing flames of fire. What would they call this spectacle?

The Star of Bethlehem (Matthew 2:1–12)

The Gospel of Matthew recounts how wise men from the East were guided to the birthplace of Jesus by a mysterious star:

- "After they had heard the king, they went on their way, and the star they had seen when it rose went ahead of them until it stopped over the place where the child was." (Matthew 2:9, NIV)

Some have proposed that this "star" was not a typical astronomical event, but perhaps a luminous object under intelligent control. While mainstream biblical and historical scholarship generally identifies the star as a comet, planetary conjunction, or some other natural event, UFO theorists sometimes cite this passage as evidence for an ancient extraterrestrial visitation. Even leaving out the UFO description, what was the bright object those men followed as it led them to Bethlehem? We truly cannot know.

The Pillar of Cloud and Fire

The Pillar of Cloud and Fire (Exodus 13:21–22): During the Israelites' exodus from Egypt, God is described as leading them by day in a pillar of cloud and by night in a pillar of fire. These unique phenomena are usually interpreted as miraculous signs. Many today, have more recently speculated about whether there could have been technical explanations. With our modern technology, could we replicate this without too much difficulty?

The Miraculous Transfer of Philip, As Documented in the New Testament

The New Testament is replete with stories of extraordinary events and individuals who were called and equipped by God to carry forth the message of the Gospel. Among these remarkable figures stands Philip, an early Christian evangelist known for his zeal, faith, and openness to the leading of the Holy Spirit.

One of the most fascinating episodes involving Philip is his miraculous "transfer" to a distant location. This sudden relocation occurred by the Spirit of God, as recounted in the Book of Acts. While any decent "Trekkie" would assume that Scotty had beamed him to his new location, I am afraid Star Trek is not often associated with biblical truths.

Who Was Philip?

Philip, often referred to as "Philip the Evangelist" or "Philip the Deacon," should be distinguished from Philip the Apostle, though both were active in the early Christian movement. Philip the Evangelist first appears in Acts 6 as one of the seven men chosen to assist the apostles in ministerial duties.

He was particularly identified for serving the Greek-speaking widows in the early Jerusalem Church. These seven, often considered the first deacons, were selected for their good reputation, wisdom, and being filled with the Holy Spirit. Philip's selection is a testament to his character and spiritual maturity.

Philip's Ministry in Samaria

After the infamous martyrdom of the preacher known as Stephen, and during the subsequent persecution of the burgeoning church in Jerusalem, many believers were scattered throughout the regions of Judea and Samaria. Philip took this as an opportunity to proclaim the Gospel in Samaria, a region historically estranged from the rest of the Jewish homeland. His ministry there was marked by powerful preaching, miraculous signs, and widespread conversions. Demons were cast out, the sick were healed, and many came to faith in Jesus by his testimony. Philip's work in Samaria set the stage for the subsequent involvement of the apostles Peter and John, who came to pray for the new believers and witnessed the outpouring of the Holy Spirit.

The Encounter on the Road to Gaza

The story of Philip's transfer is embedded within another remarkable event: his encounter with the Ethiopian eunuch. According to Acts 8:26-40, an angel of the Lord instructed Philip to go south to the road that runs from Jerusalem to Gaza. There, he encountered an Ethiopian official, a court treasurer serving under the queen of Ethiopia, who had traveled up to Jerusalem to worship and was now returning home. The eunuch was reading the book of Isaiah but struggled to understand its difficult meaning.

Philip, prompted by the Holy Spirit, approached the eunuch and asked if he understood what he was reading. The Ethiopian responded, "How can I, unless someone guides me?" (Acts 8:31). Invited to sit with the eunuch, Philip explained that the passage from Isaiah referred to Jesus. He explained about His suffering, brutal death, and glorious resurrection. The eunuch, moved by Philip's teaching and the message of the Gospel, asked to be baptized at a nearby body of water. Philip agreed and baptized him. This was in witness of the Ethiopian's joyous acceptance of faith.

The Miraculous Transfer: Philip's Sudden Relocation

It was immediately after this pivotal baptism that the New Testament records Philip's miraculous transfer. Acts 8:39-40 states: "When they came up out of the water, the Spirit of the Lord suddenly took Philip away, and the eunuch did not see him again, but went on his way rejoicing. Philip, however, appeared at Azotus and traveled about, preaching the gospel in all the towns until he reached Caesarea."

This passage describes what many commentators have considered to be a supernatural event. The Greek verb that is used here for "took Philip away" (ἥρπασεν, hērpasen) suggests a forceful or sudden removal. This is similar to the type of language used for Elijah being taken up in the Old Testament, or Paul's reference to being "caught up" to the third heaven. Philip was not merely just inspired internally. Instead, he was physically relocated (transported) by the direct intervention of the Holy Spirit. It should be specifically noted that the starship Enterprise was completely unavailable to him at the time. Scotty and his transporter could not help in this instance!

Traditional Theological Interpretations of the Bible's Tales

Most mainstream biblical scholars and theologians view these passages as deeply symbolic, rooted in the religious and cultural context of their times. They believe that the language of clouds, fire, and flying chariots primarily reflected the ancient perceptions of the divine. These scriptures were meant to convey God's awe, power, and mysterious capabilities. For many theologians, these descriptions have been understood as nothing but metaphors for God's presence and activity among humans. They have chosen to not view them as any literal accounts of His presence, and particularly not in reference to any type of advanced technology.

Considering this, we see things today that would have been completely miraculous back in those days. What would they do if we showed them the capabilities of a smart phone, if all of its apps were still functional? How would they attempt to explain an innovative flashlight? Light comes from the sun or from fires for those folks, yet this source is not hot, comes on or goes off with the flip of a switch, is able to show its light in different colors, can project a beam for more than a mile, and has a strobe capability. What if they saw a helicopter or something as large as a 747 flying, or even a semi-trailer truck moving along one of their roads? How could they possibly understand all or any of these things?

Ancient Astronaut Theory

By contrast, proponents of the "ancient astronaut" or "paleo-contact" hypothesis suggest that humanity has been visited by extraterrestrial beings throughout history, and that such encounters are preserved in myths and religious texts. They argue that certain biblical descriptions, especially those with precise technical, or mechanical characteristics, might be best explained as misunderstood sightings of alien visitors or machines. Erich von Däniken's renowned book called "Chariots of the Gods?" popularized this theory in the 20th century, though it remains highly speculative and controversial still today.

Modern Impact and Pop Culture

The idea of UFOs in the Bible has had a significant impact on popular culture, inspiring books, documentaries, and fictional works. Television programs like "Ancient Aliens" have brought these theories to wide audiences, blending entertainment with speculative history. The intersection of UFO lore and biblical narrative continues to fascinate those with an interest in both the mysteries of faith and the unknown possibilities within the universe.

Conclusion

The Bible contains numerous accounts of remarkable celestial and aerial phenomena, described in the poetic and symbolic language of their time. Whether these stories point to encounters with extraterrestrial technology, divine intervention, or the depths of human imagination, they continue to provoke wonder and inquiry.

Ultimately, the passages associated with UFOs in the Bible invite us to reflect upon the limits of our knowledge, the breadth of our curiosity, and the enduring power of ancient stories to inspire new questions in every generation. Whether regarded as evidence of visitors from the stars or as the expressions of spiritual awe, these miraculous tales remain a vibrant thread within the tapestry of human myth and faith.

❖ Exploring the Enigma of Aerial Phenomena in Vedic and Epic Traditions of India

Throughout history, the wonders and mysteries of the skies have captivated humanity's imagination. In recent centuries, reports of UFOs have sparked much debate about the presence of any extraterrestrial life and advanced technology beyond our known comprehension.

However, the fascination with unusual aerial phenomena is far from new. In the ancient Indian subcontinent, mythological and religious texts from as far back as 1500 BCE described flying chariots, celestial vehicles, and enigmatic beings that appeared to transcend the limitations of their era.

Vimanas: The Flying Chariots of the Gods

The term "Vimana" (विमान), in classical Sanskrit, literally means "measuring out" or "having been measured out." Yet in popular usage, it refers to a chariot or palace that flies through the sky. Vimanas are frequently mentioned in ancient Indian texts, and while interpretations vary, they are commonly described as flying machines or aerial chariots used by gods, heroes, and sometimes even mere mortals. So what were they?

Vimanas in the Vedas

The earliest hints of aerial vehicles appear in the Rigveda though references are less explicit than in later texts. The Vedic deities, such as Indra and Agni, are described as traversing the heavens in shining "chariots" or "cars" drawn by horses, birds, or even supernatural powers. While most recent scholars prefer to consider these as poetic metaphors, some proponents of ancient astronaut theories argue that these descriptions reflect encounters with advanced technology. Those faithful to their ancient dogmas would still argue that they are exactly what they said they are. These would be flying craft driven by their revered gods.

In the Rigveda, for example, Indra's chariot is described as "golden," "resplendent," and capable of traveling "as fast as thought." Agni, the fire deity, moves in a chariot that "illuminates the sky." These passages, although symbolic in mainstream interpretation, have been cited as early evidence of flying craft. Are these really so very different from the UAPs of our times?

Vimanas in the Ramayana

The Ramayana, attributed to the poet Valmiki, is one of India's oldest and most celebrated epics, dated to roughly 5th to 4th century BCE. The text contains one of the most detailed accounts of a flying vehicle, known as the Pushpakar Vimana.

The Pushpakar Vimana was originally owned by Kubera, the god of wealth, and later seized by the demon king Ravana. After Ravana's defeat, Rama, Sita, and Lakshmana use the Pushpakar Vimana to return to Ayodhya. The Ramayana describes the Pushpakar Vimana as a wondrous, self-propelled aerial vehicle, adorned with gems and capable of accommodating many passengers. It could travel at great speeds and was directed by thought or will.

Here is an excerpt from the English translation of the Ramayana describing the vimana:

"The Pushpaka Vimana that resembles the Sun and belongs to my brother was brought by the powerful Ravana; that aerial and excellent Vimana, going everywhere at will… that chariot resembling a bright cloud in the sky."

Does this also sound like the story of Elijah?

Some modern interpreters see both of these tales as ancient accounts of a highly technologically advanced aircraft or spacecraft. Can we prove it wasn't?

Vimanas in the Mahabharata

The Mahabharata, another monumental Indian epic, also contains intriguing references to flying chariots and mysterious aerial battles. Gods and heroes are depicted as mounting vimanas to traverse vast distances and to wage war in the sky. These are sacred texts from long ago, but could these have been UAPs, or even some of the aircraft we fly today? If they observed an F-35 stealth fighter launch an AIM-120 AMRAAM air-to-air missile against an enemy aircraft, wouldn't they describe it similarly to their descriptions within these texts?

For instance, Arjuna, one of the Pandava brothers, is granted a flying chariot by the god Indra. This chariot is said to be "powered by the mind" and capable to "go anywhere at will." There are vivid descriptions of celestial weapons, including beams of light (could these have been lasers?) that could destroy other craft. There were types of explosions that some have likened to the munitions that we use today in our modern warfare. How can we absolutely state that the writers of these tales were incorrect in their accounts, rather than to believe they actually saw what they described?

In one episode, the sage Asvathama unleashes a weapon called the Brahmastra, equated by some to be a nuclear missile, which brings "fire and smoke" and can devastate entire regions when it strikes. Such descriptions have fueled speculation about ancient advanced technologies or lost civilizations. From the Bible, does this sound similar to the sudden destruction rendered upon Sodom and Gomorrah? It seems to sound like it to me.

Other Classical Texts and Descriptions

Beyond the two great epics, numerous other Sanskrit texts and treatises make mention of flying vehicles. Some of the most notable are:

- The Samarangana Sutra Dhara: Attributed to King Bhoja (11th century CE), this treatise on architecture and engineering contains a chapter on the construction and operation of vimanas. It describes diverse types of aerial machines, the materials used, and their propulsion systems. Although the details are often seen as fanciful and not consistent with modern engineering, did the writers simply lack the technical terms to better describe these flight vehicles more realistically?

- The Vaimanika Shastra: Purportedly written in the early 20th century by Pandit Subbaraya Shastry, he claimed to receive the text through psychic channeling. This work describes in detail the design, operation, and the weaponry of vimanas. While many historians consider it as nothing but a modern forgery, it has been influential in UFO and ancient astronaut circles. It describes the utilized metals, fuels, and flight mechanisms in more technical terms. Still, it was reported to have been written before the times of widespread knowledge of jets or rockets.

Cultural Impact and Legacy

The idea of vimanas and flying chariots continues to inspire contemporary Indian art, literature, and even popular media. Movies, television shows, comic books, and video games have drawn on these ancient motifs, blending them with science fiction and modern sensibilities. Their influence will continue to affect the attitudes of their believers.

These mythological texts of India offer a treasure trove of stories about gods, heroes, and their wondrous vehicles. The vimanas, in particular, have captured the imagination of generations, both as symbols of spiritual ascent and as tantalizing hints of lost knowledge or contact with other worlds. Whether viewed as allegory or literal history, these tales speak to humanity's eternal curiosity about the skies and the mysteries that lie beyond. The dialogue between mythology and modern UFO lore is ongoing. This is surely a testament to the enduring power of revered stories and our unending quest to explore the unknown.

❖ **Exploring Ancient Sky Mysteries Through Mayan and Aztec History**

The enduring fascination with unidentified flying objects transcends many common cultures and geographic epochs. Among the most captivating associations in this topic are those linking ancient Mesoamerican civilizations, particularly the Maya and Aztec, with mysterious aerial phenomena. While contemporary UFO narratives are rooted in 20th century and 21st century popular cultures, the enigmatic art, architecture, and mythology of the Maya and Aztec have fueled a great deal of modern speculation.

Are these details about possible ancient encounters tales of unknown beings coming down from the heavens? It is time to explore the intersections of Mayan and Aztec cultures with the concept of UFOs and modern technology, considering both the historical context and our present-day interpretations.

The Mayan Civilization and Celestial Mysteries

The Maya were extraordinary astronomers. Their civilization flourished from approximately 250 to 900 CE in what is now southern Mexico, Guatemala, Belize, and Honduras. They had a profound relationship with what they perceived as their connections with the cosmos. They meticulously observed planetary cycles, solar and lunar eclipses, and charted the motion of Venus and other celestial bodies.

The construction of monumental cities, such as Chichen Itza and Tikal, was often intentionally aligned with key astronomical phenomena. These choices reflected their apparent belief in the profound connections between the heavens above and their own earthly lives below.

Glyphs and Art: Evidence of UFOs?

Some proponents of ancient astronaut theories point to Mayan carvings and codices as evidence of possible UFO sightings. Certain stelae and ceramics depict figures surrounded by what appear to be rays of light or riding upon serpent-like or skyborne creatures.

The most famous example is the sarcophagus lid of King Pakal in Palenque, which some interpret as depicting a human figure operating what looks like a rocket or spacecraft. However, mainstream archaeologists assert that this iconography is symbolic, representing the journey of the soul into the underworld, with cosmic motifs rather than literal flying machines.

Personally, I believe that if you showed a picture of this depiction to most engineers, they could identify potential elements of flight controls and life support equipment being present in this depiction.

Divine Messengers and Celestial Entities

Mayan mythology is replete with sky gods and otherworldly messengers. Deities such as Itzamna, the creator god, and Kukulkan (the feathered serpent), were associated with the sky and cosmic order. Stories of these gods descending from or ascending to the heavens are extremely common, often involving dramatic celestial events. While some believers interpret these narratives as potential records of UFO encounters, they are generally understood by mainstream scholars and skeptics to be within their cultural context of myth, ritual, and their understood cosmology.

The Aztec Empire: Omens and Heavenly Phenomena

Aztec Cosmology and Portents

The Aztec civilization reached its zenith in the 15th and early 16th centuries in central Mexico. They placed immense importance on celestial events. The Aztec calendar, temples, and key societal rituals were closely tied to their observations about the motions of the sun, moon, and stars.

Aztec priests interpreted unusual phenomena in the sky, including comets, eclipses, and meteor showers, as omens from the gods. These omens were believed to signal significant world changes or even impending disasters. Such celestial signs were held in high regard by these ancient societies.

Codices and Reports of Strange Sights

Several Aztec written codices and subsequent Spanish colonial chronicles record unusual events that were also witnessed in the skies before the arrival of the early conquistadors and their later Spanish settlers. Notably, Bernardino de Sahagún's "Florentine Codex" describes "a great flame in the sky" and "blazing fireballs" in the years leading up to the conquest of Tenochtitlan. These were seen as dire portents by the Aztecs, who believed that the gods communicated through such signs.

While some modern enthusiasts equate these descriptions with UFOs, most historians interpret them as references to natural phenomena such as meteors, comets, or volcanic activity. Could they have been both?

Gods from Above: The Role of Quetzalcoatl

The Aztec feathered serpent god, Quetzalcoatl, shares similarities with the Mayan Kukulkan. Quetzalcoatl was believed to have arrived from the east, being often associated with the planet Venus and various celestial cycles.

Some fringe theories suggest that the veneration of sky or star-related deities could be rooted in misunderstood extraterrestrial contact. However, these interpretations frequently lack any significant support among most scholars of Mesoamerican religions and myths.

Does this immediately disqualify these myths as false? Learned scholars do not tend to believe these similar tales, even while they are told by differing civilizations, often widely varied in both timeframe and geographic locations.

❖ The Star People in Anasazi Native American Lore

The Anasazi, also known as the Ancestral Puebloans, were an ancient Native American civilization that inhabited the Four Corners region of the United States, where Colorado, Utah, Arizona, and New Mexico meet. Renowned for their remarkable stone dwellings, kivas, and cliff-side cities like Chaco Canyon and Mesa Verde, the Anasazi have left behind not only fascinating ruins but also a rich legacy of myth and oral tradition. Among their many legends that have survived across generations, sincere tales of the "Star People" stand out as some of the most intriguing and enigmatic ones even to our day.

Let us explore the concept of the Star People within the perspective of the early Anasazi culture. To do this, we need to examine the traditional beliefs, oral legends, and archaeological evidence that has been kept within the societies of their descendants. Doing so may allow us to comprehend the broader significance of these stories for understanding the worldview held by these Ancestral Puebloans.

Who Were These Anasazi Tribes?

The term "Anasazi" is not clearly known to be the names these people would have used to describe themselves. Instead, this term is derived from a Navajo word that roughly translates to "Ancient Ones" or "Ancient Enemy." Today, modern successors of the Anasazi lands and history prefer to use the term "Ancestral Puebloans," which acknowledges their continuity with modern Pueblo tribes such as the Hopi, Zuni, and various Rio Grande pueblos.

The Anasazi civilization is credited as flourishing from approximately AD 100 to the year 1600. They were obviously skilled architects, farmers, and astronomers, constructing vast networks of roads, ceremonial centers, and astronomical observatories. Their geometrically spoked roads and alignments to celestial events suggest a deep engagement with the sky, which is further reflected within their detailed myths and spiritual beliefs.

The Concept of the Star People

The idea of "Star People" refers to beings believed to have come down from the heavens to Earth. They are often associated with stars, particular star systems, or other celestial phenomena. This concept is found in various indigenous cultures across North America, certainly including those descended from the Anasazi themselves. These stories, handed down generationally through careful oral traditions, are not typically found in any written form but are instead woven into the ceremonial and spiritual fabric of these people.

Legends and Oral Traditions

Many Pueblo stories recount important encounters with mysterious beings who descended from the sky in beams of light or who travelled across the heavens in luminous flying vehicles. In some versions of these legends, the Star People are described as tall, luminous, and were unlike any humans known to their cultures. They are said to have shared great wisdom, taught the people how to read the stars, and instructed them in the building of ceremonial sites that were aligned with celestial events.

One common motif in these tales is the "Sky Kachinas," or spiritual beings, who function as their intermediaries between the celestial worlds and the Earth. The Hopi, in particular, have a complex pantheon of Kachinas, some of whom are said to come from the "star cluster" or other celestial realms. While not all Kachinas are explicitly "Star people," there are specific beings in their mythology who are associated with notable constellations, planets, and stellar phenomena.

Symbolism in Art and Architecture

The Anasazi left behind a wealth of petroglyphs and pictographs, many of which depict celestial objects like suns, moons, stars, and enigmatic figures with radiating lines or unusual headgear. In Chaco Canyon, for example, rock art shows what some scientists may interpret as supernovae or comets, and mysterious humanoid forms that some theorists link to these star beings. A fair number of them even depict what some believers consider to be early astronauts, even including what may appear to be spacesuits and helmets.

Their great kivas and ceremonial plazas are often aligned with solar and lunar events, suggesting that the sky was central to their belief system and daily life. Archaeologists have found that certain buildings mark solstices and equinoxes with remarkable precision. There is supporting evidence of quite sophisticated astronomical knowledge demonstrated. Their descendants have attributed these to the very teachings revealed by the Star People within ancient legends.

Comparisons with Other Indigenous Star Beings

The concept of Star People is not unique to the Anasazi and Puebloans. Similar tales are found among the Lakota (Sioux), who speak of the "Star Nation;" the Cree and Ojibwe, who honor the "Sky People;" and the Navajo, who have complex myths involving the constellations and holy beings who have descended from space unto our Earth. These stories often serve to explain the origins of the diverse peoples and tribes, the understood structure of this world, and the relationship between humanity and the cosmos.

Celestial Knowledge and Its Significance

For the Anasazi and their descendants, the sky was a living text, filled with meaning and guidance. The cycles of the sun, moon, and stars governed planting and harvesting seasons, the timing of ritual ceremonies, and directed the rhythms of community life. By encoding this knowledge in myth, art, and architecture, the Anasazi ensured its transmission would be successfully carried across the many generations to come.

The Star People, therefore, whether viewed as literal visitors or as symbolic representations of celestial wisdom, embody the reverence and curiosity with which the Anasazi approached the grand cosmos. The legends emphasize the interconnectedness of all things and the belief that the universe is animated by conscious, guiding forces.

Modern Perspectives and Cultural Revival

The mysteries of the Anasazi, including their abrupt departure from certain sites and the purpose of their most enigmatic structures, continue to inspire both scholarly research and popular imagination. The famed stories of the Star People persist today as a strong testament to the Anasazi's enduring fascination with the heavens and serve as a reminder of the depth and richness of indigenous cosmologies.

Whether in South America, Central America, or North America, their history continues to intrigue the people of today. This is largely because of the shared natures of these ancient tribes, their joint fascination with the stars and heavenly observations, and their common stories of gods who came down to the Earth from the stars.

Could they all have been so completely mistaken in their convictions?

Yet Another of My Own Stories:

Search and rescue missions, or SARs, were a significant part of the duties for a flight surgeon located at Patrick AFB while I was there. These were always important because some person, or a group of people, had their lives dependent on whether we could rescue them. The vast majority of these cases for us were out into the Atlantic Ocean.

The circumstances of the SAR missions could vary widely. Some were boats that sank or had become disabled while others were related to aircraft that had landed in the ocean. One case, however, was far from the normal situation for my aircrew and me.

We received word that there was a Chi-Com (Chinese Communist) cargo vessel that had radioed in that they had a sailor who was ill and needed immediate evacuation to a hospital for emergency surgery. We were briefed before take-off that this ship had a helipad suitable to receive one of our large HH-3E Jolly Green Giant helicopters, so we headed out to perform our mission.

First, though, allow me to explain about why our helicopters had such a designation. The legendary "Jolly Green Giant" helicopters were the legacy of the Sikorsky HH-3E and its successors. This would later include the much larger Sikorsky HH-53B/C and MH-53 Pave Low helicopters, that came to be known as the "Super Jolly Green Giants."

The original "Jolly Green Giant" helicopters are an enduring symbol of hope, bravery, and technical ingenuity within the annals of military aviation. Best known for their pivotal role in combat search and rescue

(CSAR) missions during the Vietnam War, these jungle green giants have saved countless lives and forged a unique legacy in U.S. Air Force history.

The story of the Jolly Green Giant helicopters began in the late 1950s as the U.S. military sought a reliable, all-weather search and rescue aircraft. The Sikorsky Aircraft Corporation answered the call, building on the success of its S-61/SH-3 Sea King. This twin-engined, amphibious helicopter was known for anti-submarine warfare in the U.S. Navy. The U.S. Air Force needed a similar stout platform capable of deep-jungle and other combat penetration rescues, and Sikorsky adapted their prior designs, leading to the birth of the HH-3E.

The HH-3E was nicknamed as the "Jolly Green Giant" by those who flew and maintained it. They had to receive special permission from the vegetable producing company who owned the patent for this moniker. One curious fad used by the flight crews who manned it was to use the Giant's footprints, a pair of green feet, on units' logos, on the hangar doors of the units, and even in tattoos on the person of crewmembers and support staff. It was also not uncommon for Jolly crews to paint green feet on the hangars and taxiways of flight units they visited.

I recalled once landing at a Coast Guard airbase in the Bahamas and as we landed, the crew was forbidden by their security personnel from egressing the helicopter. The commander personally came to our crew and forcefully directed that there would be no painting of green feet anywhere in his hangar or flightline. It seemed that a previous Jolly crew had left behind more than fifty sets of green feet all over their hangar, and he would not allow us to enter his building until we had promised, as proper military personnel, that no feet painting were be perpetrated on his installation on this occasion. I can neither confirm nor deny that more than a dozen pairs of green feet had mysteriously generated themselves at the time of our departure the next day.

The HH-3E was designed from the get-go for survivability and endurance in hostile environments. It boasted some unique innovations for its time:

- Self-sealing fuel tanks were used that could withstand penetration by small-arms fire
- Armor plating was added for the protection of the crew
- External fuel tanks were added to allow for an extended range
- Its rescue hoist was fitted with a 240-foot cable and jungle penetrator for lifting survivors up from difficult terrain and jungle environments
- An in-flight refueling probe, one of the first helicopters equipped with this feature, helped to enable unprecedented range of flight operations. The most common refueler used by the Jollies was the renowned HC-130. This pair of aircraft were created in search and rescue heaven.
- The HH-3E was designed with an amphibious hull, allowing for water landings and take-offs. The trick to this was that the heavy engine turbines on the top of the fuselage made it unsteady in rough seas, with a few cases where the aircraft rolled over into the water. Not a good day for anyone aboard!

The HH-3Es generally sported an olive-drab color when used in jungle environments. Along with its hulking size, and its life-saving missions, this was why it earned the affectionate label and logo of the "Jolly Green Giant," nicknamed after the familiar mascot of a popular vegetable brand by the same name.

The Vietnam War was the crucible of combat that made the Jolly Green Giant famous. This war placed extraordinary demands on search and rescue operations, due to the nature of the local jungle topography itself.

Downed pilots and seriously wounded soldiers often found themselves deep behind enemy lines or within perilous terrain, where time was of the essence. The U.S. Air Force's Aerospace Rescue and Recovery Service (ARRS) deployed the HH-3E as its workhorse, making it synonymous with courage under fire.

Flying thousands of Combat Search and Rescue (CSAR) missions, this flight platform became renowned. When American pilots were shot down during bombing raids over South Vietnam, North Vietnam, Laos, and even Cambodia, it was up to the HH-3 crews to find and rescue them.

Each mission required extreme bravery from the pilots, flight engineers, pararescuemen (PJs), and gunners. They often faced anti-aircraft fire, dense jungles, and the ever-present threat of enemy forces converging on the rescue site. Every mission was filled with danger.

Notable features that set these missions apart included:

- Jungle insertion of pararescuemen by hoist or rope into hostile terrain. Recovery of these airmen and their rescued personnel might occur using the jungle penetrator or by using a Stokes litter.

- Mid-air refueling from the HC-130 tankers, which allowed deep penetration into enemy territory.

- Escorts were often needed from heavily armed Huey and AH-1 Cobra helicopters. They also depended on fixed winged aircraft, such as the A-1 Skyraider attack planes, providing suppressive fire as the helicopters hovered over many deadly scenes.

The heroic deeds of Jolly Green Giant crews became the stuff of legend. During the war, they retrieved thousands of downed airmen and injured soldiers. Their valor was recognized with numerous awards, including Air Force Crosses and Medals of Honor. Some of the most famous PJs, such as Air Force pararescue legend William H. Pitsenbarger, conducted their missions with a mix of technical skill and unwavering resolve.

As the Vietnam conflict intensified, the U.S. Air Force sought for greater capabilities. Enter the Sikorsky HH-53B/C, soon known as the "Super Jolly Green Giant." Larger, faster, and more heavily armed, the HH-53 could carry more survivors, operate at greater ranges, and withstand heavier fire than its older brother. Its twin turbine engines, well-armored cockpit, and expanded ramp-equipped fuselage represented major advances. The "Super Jolly Green" and its crews became the second backbone of CSAR in Southeast Asia.

They operated alongside the HH-3Es through the end of the conflict. Their role did not end with Vietnam, as they flew, for instance, in the hazardous Mayagüez Incident in 1975. They continued to serve in a variety of rescue, special operations, and humanitarian missions across the world, long after the war in Southeast Asia had ended.

The MH-53 Pave Low helicopters were more specialized for the many special operations missions they supported around the globe. In this role, the Pave Low missions often required low-level, long-range, and undetected penetration actions into enemy areas. Their revered ability to fly day or night, to remain stable in unfavorable weather, gave these Sikorski helicopters the capability to allow for insertion, egress, and even key resupply missions for a variety of special operations forces.

After the Vietnam War, the Jolly Green Giant helicopters continued to evolve. My crews and I used the HH-3E for SAR missions and space shuttle support until it was retired from active service in the 1990s. It was then replaced by more advanced aircraft, especially the Sikorsky HH-60G Pave Hawk. I was fortunate enough to fly missions in both of these aircraft while stationed at Patrick AFB and Cape Canaveral AFS. Nevertheless, the spirit of the Jolly Green Giant persists in today's rescue squadrons, whose members still wear the green feet and uphold the famed PJ/Rescue Service motto: "That Others May Live."

Getting back to the mission we were to fly to the Chi-Com freighter, everything seemed perfectly fine as we flew east from Patrick and out over the Atlantic. We were correctly guided to the ship with the assistance of ground radar systems. It was a large ship as we approached, and the big helipad was well marked. Believing all to be well, our pilots guided the large helicopter downward and we made an uneventful landing on the venturing freighter.

That was when the circumstances suddenly changed. We saw a squad of armed men, probably more than a dozen of them, who ran up onto the helipad and surrounded our aircraft. Inside, we were worried that we had perhaps flown into some sort of trap conjured up by the Chinese Communists. Then, we watched what happened next.

The circle of armed guards turned outward, pointing their rifles out toward the other crewmen. We quickly realized that this was a self-protective measure by the Communist sailors. If any of their crew tried to jump aboard our helicopter, as an attempt to gain political asylum, those AK-47s would be used to mow down any of the Chinese crewmembers who had such thoughts.

The ill crewman was soon brought to our aircraft, and we were cleared to take off. We did this immediately!

It turned out the involved sailor had appendicitis, and we flew him to a local hospital in Melbourne, Florida, where he had his surgery. He was then given back to the representatives of his government. In the end, there was no international incident.

I will say, however, that when an American aircrew lands on a ship to provide life-saving care, it isn't at all pleasant to be surrounded by a group of men carrying automatic weapons!

We Got it From Them!

Chapter 3

Modern Interpretations and Ancient Astronaut Theories

❖ **It Is in Our Popular Culture, but is it Just Another "Pseudoscience"?**

The 20th century saw a surge of interest in the idea that ancient monuments and myths are evidence of extraterrestrial visitation. Books like Erich von Däniken's "Chariots of the Gods?" popularized the notion that the Maya and Aztec, among other civilizations, may have had contact with aliens. Images of Mayan artifacts, allegedly depicting astronauts or spaceships, circulated widely online, feeding into modern UFO lore.

Critical Analysis

Archaeologists and historians overwhelmingly reject these theories as pseudoscientific. They point out that Mesoamerican art and myth are best understood through the lens of indigenous cosmology, ritual, and symbolism. The apparent "modernity" of some designs, such as the depiction of beings with ornate headdresses (helmets), or sky chariots, are trusted to be consistent with traditional religious iconography.

One argument I have about this is that I have seen some of these "birds" shown on episodes on this subject. What archaeologists claim to be simple avian figures absolutely look like aircraft. If one of these archaeologists took one of these depictions to any grade school in the country and asked the children what the figures represent, the kids will universally say that these are airplanes rather than birds. The figures have two wings, a fuselage, and an obvious tail. They absolutely look like airplanes. It is curious that school kids can easily recognize the figures when trained academics cannot!

Why Do UFO Theories Persist In Reference to The Ancient Cultures?

The appeal of connecting ancient civilizations with extraterrestrial visitors is deeply rooted in both the mystery of lost knowledge and a fundamental human fascination with the unknown. In the case of the Maya and Aztec, their sophisticated understanding of astronomy, mathematics, and architecture has inspired awe and curiosity for centuries. This is as we have already discussed. Modern UFO interpretations often reflect the hopes, fears, and imaginations of the present, projected onto the enigmatic past. But is this appropriate?

There is no credible historical or archaeological evidence to absolutely suggest that the Maya or Aztec peoples experienced or recorded encounters with UFOs in the sense understood today. This is also true for cultures in India, Egypt, Australia, and for the descendants of the Anasazi of North America. Somehow, there seems to be no photographs available.

Yet the legacy of the diverse ancient people's fascination with the heavens continues to inspire our wonder today. Those people truly believed they met beings from the heavens. Their various stories, art, myths, and architecture reflect a worldview in which the sky was alive with meaning. They had firm beliefs in realms inhabited by celestial gods, sky omens, and cosmic cycles. The enduring allure of ancient UFO mysteries tells us as much about our own desire to connect with the unknown as it does about the distant civilizations themselves.

In the end, the real miracle may be the creativity and curiosity with which humans, past and present, gaze upwards, always seeking answers in the infinite canvas of the sky above us.

❖ The American Government's Involvement and Declassified Files

We know that the U.S. Air Force initiated Project Blue Book in 1952. Its stated goal was to investigate the range of UFO reports from across the country. Over nearly two decades, Blue Book collected and analyzed more than 12,000 cases, ultimately concluding that most sightings were attributable to natural or man-made phenomena. Still, there was a small percentage, of about 6%, that remained unexplained.

Declassified files from other countries, including the United Kingdom, France, Argentina, Spain, and Russia, reveal that they have undertaken similar efforts to track, identify, and analyze these unexplained UAP reports. It is not at all limited to the U.S.!

AARO and Government Reports Concerning UAP Evidence

In recent years, public interest in UAPs has surged, fueled by an erratic series of official government disclosures and reports. Quite often, one report may diametrically oppose one from a different agency. This is a cause of endless frustration within the UFO community as they beg for greater transparency from our government.

At the forefront of the United States' contemporary effort to study these phenomena is the All-domain Anomaly Resolution Office (AARO). This is an official Department of Defense body tasked with investigating reports of UAPs across multiple domains: aerial, maritime, and space.

This analysis will explore the formation, purpose, and findings of AARO within the broader context of previous governmental reports on UFO evidence. I hope to present a comprehensive overview of what is known, what details remain unresolved, and what the implications are for our national security and scientific inquiry.

❖ The Evolution of U.S. Government Interest in UFOs

The U.S. government's involvement with UFO investigations dates back to the late 1940s, beginning with the U.S. Air Force's Project Sign (1947), followed by Project Grudge (1949), and the more widely known Project Blue Book (1952–1969). These efforts aimed to determine whether UFOs posed a threat to national security and to scientifically analyze UFO-related data. Project Blue Book investigated thousands of sightings, most of which were attributed to misidentified natural phenomena or human-made objects. However, a small percentage remained unexplained.

Upon the termination of Project Blue Book in 1969, the Air Force concluded that UFOs did not represent a threat to national security, had not exhibited advanced technology beyond current human understanding, and that there was no evidence to suggest any form of craft having any extraterrestrial origins. However, the public was not convinced.

The Resurgence of Official Interest: The 21st Century

For decades, UFO research languished on the periphery of official defense and scientific initiatives. However, renewed interest emerged in the 21st century. In 2017, The New York Times published articles revealing the existence of the Advanced Aerospace Threat Identification Program (AATIP). This was revealed to be a secret initiative within the Department of Defense that investigated UAP reports between 2007 and 2012. This disclosure sparked a wave of public and legislative calls for greater transparency.

In response, the Pentagon declassified and released several Navy F/A-18 tactical videos showing unidentified aerial objects exhibiting flight characteristics beyond known technologies. These videos, commonly referred to as "FLIR1," "GIMBAL," and "GOFAST," prompted further scrutiny from lawmakers, scientists, and the public in general.

The Creation and Mission of AARO

In July 2022, the Department of Defense formally established the All-domain Anomaly Resolution Office (AARO) to coordinate efforts across military and intelligence agencies to investigate UAPs. The creation of AARO followed provisions in the National Defense Authorization Act (NDAA) for Fiscal Year 2022. The provision, which mandated a centralized office dedicated to addressing UAPs, did so as a matter of national security.

AARO's mission was supposed to encompass the UAP detection and identification. They were to study objects of interest in, on, or near military installations, their operating areas, their training areas, their strategic storage facilities, and their special-use military airspace.

In essence, they had been tasked to examine all areas of national security concerns. The office was also tasked with assessing and mitigating any associated threats to the safety of civilian operations and overall national security. This was supposed to be the military agency that would be prepared to solve this UAP matter once and for all.

Mandate and Reporting Structure

AARO collaborates with various defense, intelligence, and scientific entities, including the Office of the Director of National Intelligence (ODNI), NASA, and the Federal Aviation Administration (FAA). It maintains a centralized repository for UAP reports and is responsible for delivering regular classified and unclassified briefings to Congress. Its reach was expected to be dealing with any of these anomalous cases using the entire technical capabilities of our government.

The 2021 Preliminary Assessment

In June of 2021, the Office of the Director of National Intelligence released a preliminary report examining UAP data collected by military personnel between 2004 and 2021. The ODNI report reviewed 144 specific incidents, with investigators able to explain only a single case. This was found to be a deflating balloon. The remaining 143 cases remained largely unsolved, with some exhibiting unusual flight characteristics. These included high velocity craft, the lack of discernible propulsion systems, the lack of traditional flight control surfaces, and the ability to rapidly change direction. This latter finding included incidents that were noted to exhibit high G-force loads on the craft, possibly over 1,000-Gs, which could not be tolerated by any conventional aircraft.

However, the report noted that they had found limited amounts of high-quality data and reported that this finding impeded definitive conclusions from within the Pentagon. While some UAPs could represent possible technologies deployed by adversarial nations (such as Russia and China), the report found no evidence of extraterrestrial origins for these unexplained cases.

Still, they could not completely rule out extraterrestrial etiologies for these anomalous vehicles. In other words, they basically said that they don't know what these were. Despite them being totally unable to identify the origins of these extraordinary craft, the report concluded they still could not be extraterrestrial. Why?

Just because!

That was sure a big help!

Subsequent Annual Reports

AARO and ODNI have since produced annual UAP reports as mandated by Congress. These reports summarized new sightings, evaluated technological and adversarial threats, and reviewed their progress in establishing the capabilities to create standardized reporting and analysis protocols. This was important because they wanted to increase the militaries, and the general public's, ability to tell them what was being witnessed by people all across the country.

The unclassified 2022 Annual Report, for example, highlighted an increase in overall UAP reporting. This, however, was mostly believed to be due to having promoted a reduction in the stigma of reporting. Also, there had been the implementation of a set of streamlined reporting mechanisms, which made it easier to create UAP testimonies.

The annual report affirmed that most new cases were likely attributable to the same old mundane sources that had been repeatedly described in the past. These included such factors as weather balloons, drones, unusual cloud formations, misidentification of planets and stars, or airborne clutter. Despite this, as noted so frequently before, a small percentage of cases continued to remain unexplained, and under further investigation.

.

❖ Congressional Hearings and Public Engagement

Congress has held several hearings on UAPs, with military officials and intelligence representatives providing extensive testimony. Lawmakers from both parties have expressed their concern about insufficient data-sharing, the risks posed by unidentified aerial objects near sensitive sites, and the need for enhanced scientific analysis.

These hearings have therefore prompted increasing calls for greater transparency and for the release of additional data to the public. While most detailed information remained classified, the unclassified reports have helped demystify government processes related to UAPs and brought a perceived measure of increased accountability.

Persisting Scientific and Security Implications Severely Challenge the Government in Both Data Collection and Its Analysis

Investigations of UAPs have presented numerous challenges that are seemingly difficult to overcome. Many of the reports are based on eyewitness accounts, radar tracks, and a variety of sensor data that widely vary in quality. The officials related that environmental factors, sensor errors, and public misperceptions can all contribute to misidentification of known causes.

The standardization of reporting elements and difficulties in validating data remains a fundamental priority for AARO. The key organization does not seem to be capable of fully explaining the most intriguing cases. Still, the most compelling data remains classified, limiting the ability of independent scientific reviews.

Possible Scientific Explanations and Hypotheses

AARO and government reports categorize UAPs into several explanatory domains:

- Airborne clutter: These cases would include flights of birds, unknown balloons, recreational unmanned aerial vehicles, or even airborne debris.

- Atmospheric phenomena: Natural events within clouds would include the presence of ice crystals, temperature inversions, unusual regions of extreme turbulence, lenticular clouds, or other types of weather anomalies.

- U.S. government or industry developmental programs: Classified, undisclosed technology research and testing of undeclared advanced technology craft would be included here. The problem is that these items are likely to be so secret that they cannot be discussed.

- Adversary systems: Foreign versions of high technology aircraft, drones, or other surveillance systems might be deployed by rivals. Once more, the articles that are discovered in this category are unlikely to be discussed openly.

- Other: Incidents that lack sufficient data for clear attribution, including the possibility, however remote, of actual non-human origins. This is the category that would be used only under the most unusual of circumstances. After all, the government has noticeably decided that no data can be officially recognized as being of extraterrestrial origin. Duh!

It is this final category that continues to intrigue both the public and the scientific community. The possibility of disclosure of UAP events by the government would be near-zero, in my opinion. Even if it landed on the White House lawn, no public release of UAP information would likely be forthcoming.

Key Implications for Technology and National Security

Even in the defined absence of evidence for any extraterrestrial technology, the existence of unexplained UAPs near our sensitive military installations or within restricted airspace sites raises entirely valid concerns to most people. Such mysterious objects could represent advanced adversarial technologies or their intelligence-gathering platforms. Either case could pose a serious threat to operational security and to America's national defense.

NASA's Independent UAP Study

In 2022, NASA initiated an independent study of UAPs, emphasizing the importance of rigorous scientific inquiry and increased transparency. The agency convened a panel of skilled experts in aeronautics, data analysis, and communications to evaluate the existing UAP data and recommend best practices for future research. The public has largely taken a wait and see attitude about yet another governmental agency investigation.

Will NASA be able to provide additional information compared to those inquiries of the past? We cannot say that they can easily answer this question. Both lawmakers and the American public continue to demand greater access to UAP information. While operational security and intelligence considerations will always require some restrictions, the momentum for more open scientific investigation is distinctly growing. AARO and other government agencies are under increasing pressure to declassify data and to engage with academic and private sector researchers. It may be only under the directions of President Donald Trump that disclosures will finally occur. Who knows?

❖ The Results of Recent Congressional Hearings and UAP Witness Testimony

In recent years, the topic of UAPs have been vaulted from the fringes of conspiracy and speculation and into the center of national and scientific discourse in the United States. This shift has been catalyzed in part by a series of congressional hearings and high-profile testimonies from established experts from within the Department of Defense and other agencies.

These respected witnesses have brought new visibility and urgency to questions that have long lingered in the public imagination: What are UAPs? How are they being investigated? And what do they mean for national security, scientific inquiry, and public awareness?

The Catalyst: Declassified Military Footage and Reports

A significant turning point in mainstream attention to UAPs occurred in 2017. It was then that the New York Times reporters Helene Cooper, Ralph Blumenthal, and Leslie Kean, published a story revealing the existence of a clandestine Pentagon program called the Advanced Aerospace Threat Identification Program, or AATIP.

AATIP was originally begun in 2007, with a reported budget of $22 million. This was largely accomplished because of Senate Majority Leader Harry Reid, representing Nevada, Senator Ted Stevens of-Alaska, and Senator Daniel Inouye from Hawaii. Once it was up and running, a major contract was given to Bigelow Aerospace Advanced Space Studies. Robert Bigelow owned this company and was an old friend of Senator Reid.

Through this contract, Bigelow Aerospace received most of the money allocated for the Pentagon program.

A former enlisted special operations functionary with a chest full of medals, Luis Elizondo, was selected as the reported head of this AATIP operation. Curiously, he was the son of a Cuban expatriate who had volunteered for Brigade 2506. This CIA-sponsored faction of exiles was formed in 1960. This trained brigade was the very group that was later involved in the infamous April 1961 action known as the Bay of Pigs invasion.

Against terrific Pentagon headwinds, Elizondo was eventually able to get the previously unknown F/A-18 videos declassified and released. These extraordinary new videos showed strange objects maneuvering in ways that defied known aeronautical technology. The authenticity of these unique videos was later confirmed by the Department of Defense in 2020. In doing so, the Pentagon sparked widespread calls for more transparency and further investigations.

Analysis of these Released F/A-18 Super Hornet UAP Videos

In recent years, videos captured by F/A-18 fighter jets of the United States Navy have reignited global interest in the subject of UAPs. These videos, commonly known by nicknames such as the "FLIR1" (also called the "Tic Tac"), "GIMBAL," and "GOFAST," were declassified and released to the public because of AATIP and the continued efforts of Mr. Elizondo. The release of these videos prompted widespread speculation about even more videos that were still being hidden from the American public.

Despite these government admissions, there was a renewed scientific curiosity regarding what these objects might represent. Their additional analyses aimed to critically examine the content, context, and implications of the amazing F/A-18 UAP videos. In doing so, the subsequent congressional hearings sought to integrate existing expert commentary, the release of additional technical data, and to increase the possible cultural impacts created by these unexpected controversies.

Background: The Videos and Their Release

The three most significant F/A-18 UFO videos were taken between 2004 and 2015. The footage was captured using advanced forward-looking infrared (FLIR) targeting pods attached to the F/A-18 Super Hornets. The videos feature the voices of aviators and radar operators expressing astonishment at the performance and appearance of the observed objects.

- FLIR1 (also known commonly as the "Tic Tac"): Recorded in 2004 off the coast of Southern California by pilots assigned to the Nimitz Carrier Strike Group. It shows a white, elongated object maneuvering in ways that seem to defy conventional aeronautics. We will later discuss this further.

- GIMBAL: Captured in 2015 by Navy pilots over the Atlantic Ocean. The video displays a spinning, disk-shaped object that appears to move against high-altitude winds, with pilots commenting on its unusual rotation and lack of visible propulsion. The details of this encounter makes it ever more interesting.

- GOFAST: Also taken in 2015, this video depicts a small, fast-moving object flying low over the water, with the onboard sensors struggling to lock onto it. As with the others, we saw flight characteristics that cannot be duplicated by conventional aircraft.

The videos were officially released by the U.S. Department of Defense in April 2020, although they had previously been leaked and circulated in the public domain. This unprecedented move was accompanied by statements indicating the government's interest in understanding the nature and origins of these phenomena.

Technical Analysis of the Videos

The F/A-18's AN/ASQ-228 Advanced Targeting Forward-Looking Infrared (ATFLIR) pod is a sophisticated system capable of tracking airborne and surface objects using infrared imagery and laser designation. Despite its cutting-edge capabilities, interpreting ATFLIR footage can be challenging due to optical artifacts, sensor limitations, and environmental effects.

There were also limitations to infrared sensing capabilities. The objects often appear as featureless blobs due to the nature of IR imaging, which detects differences in temperatures rather than visible surface details.

Sensor locks have also been encountered. The "lock" messages and dynamic overlays on the video provide data about speed, altitude, azimuth, and more. However, interpreting these figures requires expertise, and misreading them can lead to overestimation of the objects' speed or maneuverability. This can be exacerbated by camera movements. In several instances, the apparent motion of the objects are partially attributable to the movement of the jet and the targeting pod itself.

Aerodynamic and Kinematic Observations

The pilots' reactions, captured in real time, underscore the seemingly extraordinary performance of the UAPs. The "Tic Tac" object allegedly accelerated nearly instantaneously, then changed directions rapidly. Besides all of this, it was able to hover with no visible means of propulsion. The voices of the pilots can be heard as they laugh and try to figure out whatever it was that they were seeing.

The "GIMBAL" object's apparent rotation and resistance to high winds appeared inconsistent with known aerodynamic principles, according to certain experts. Similarly, the "GOFAST" object's speed and proximity to the water raised some technical questions about its possible propulsion and control mechanisms. However, skeptical analysts, as always, have pointed out the potential for gross misinterpretation of these videos.

The Aviators' Credibility and Reporting

The credibility of the military pilots involved is a crucial factor in assessing the significance of these encounters. Their range of extensive training and experience in identifying both friendly and adversarial aircraft lends significant weight to their professional observations. Several aviators, such as Commander David Fravor, who participated in the "Tic Tac" encounter, have publicly stated that what they saw and tracked definitely defied any conventional explanation.

Based on my own military experience and training, I must state that I found the videos extremely compelling. While the visible evidence on the videos are impressive, what captured my attention was the tones of the voices from the naval aviators (who are self-determinedly better than mere Air Force pilots). These skilled flyers are highly-trained and very professional.

Yet when I hear the clear astonishment in their voices, I recognize that they are witnessing something totally different from their normal encounters. They had their own Mark-one eyeballs on those vehicles and fully recognized the anomalous craft for what they were: unexplainable for any human technology!

Public Fascination and Media Coverage

The release of these F/A-18 UAP videos has had a profound impact on public discourse and media coverage. Mainstream outlets covered the Pentagon's confirmation of the videos' authenticity, and documentaries, podcasts, and social media discussions proliferated.

The long-standing stigma that has been commonly associated with UAP reporting has decidedly lessened, as mentioned before. This has undoubtedly resulted in more military and civilian pilots coming forward with their own accounts. Their personal testimonies before Congress have had a lasting impact.

❖ Congressional Hearings: Bringing UAPs Into the Light

The 2022 House Intelligence Subcommittee Hearing

On May 17, 2022, the U.S. House of Representatives held its first open congressional hearing on UAPs in more than 50 years. Led by the House of Representatives' Intelligence Counterterrorism, Counterintelligence, and Counterproliferation Subcommittee, the hearing featured testimony from experts in many fields. There was evidence from Ronald S. Moultrie, the Under Secretary of Defense for Intelligence and Security, along with that of Scott W. Bray, the Deputy Director of Naval Intelligence.

Lawmakers were briefed on over 400 special reports of UAPs, many submitted by military aviators. Officials explained that while most incidents remained unexplained, some could be attributed to airborne clutter, natural atmospheric phenomena, or foreign surveillance. Nevertheless, a substantial number of these cases exhibited flight characteristics or technological capabilities beyond known U.S. or adversary inventories.

As with the instances of released F/A-18 videos, these kinds of authoritative explanations were seen as highly important. The officials emphasized the need to destigmatize reporting, improve data collection, and ensure that potential national security threats are properly assessed.

The July 2023 House Oversight Committee Hearing

A pivotal moment in the recent discourse came on July 26, 2023, when the House Oversight Committee's National Security Subcommittee convened a high-profile hearing on UAPs. This session drew testimony from three notable witnesses:

- David Grusch: A former Air Force intelligence officer who is a whistleblower. He claimed that the U.S. government possesses non-human "biologics" and has engaged in a multi-decade program to retrieve and reverse-engineer crashed UAPs.

  Grusch asserted, under oath, that he had knowledge of individuals harmed during efforts to conceal this information, though he stated he could provide details only in a classified setting.

  He further claimed to know military officers who had personally identified reverse-engineered craft from their own experience. The down-side to this was that none of these individuals were willing to come forward to relate their tales.

- Ryan Graves: A former Navy fighter pilot and executive director of Americans for Safe Aerospace. Graves described routine encounters with UAPs by Navy aviators off the Atlantic coast, including objects with no visible propulsion and the ability to remain stationary in high winds. He called for improved mechanisms for reporting and analyzing such incidents.

- David Fravor: A retired Navy commander and aviator who famously encountered the "Tic Tac" UAP off the coast of California in 2004. Fravor compellingly recounted the object's extraordinary acceleration and flight capabilities, which he asserted were far beyond the reach of any known technology.

These testimonies electrified public interest and led to bipartisan calls for greater transparency, robust whistleblower protections, and the declassification of UAP-related records. Plainly, these stories had significant national security implications.

A consistent theme throughout the hearings was the potential threat UAPs pose to U.S. airspace security. Lawmakers and military officials expressed concern that some phenomena could represent advanced technology from foreign adversaries, such as China or Russia, or unknown actors.

The possibility that UAPs could interfere with military operations or reconnaissance activities has driven the push for more systematic data collection and interagency coordination. Even for commercial aviation, the threats for possible mid-air collisions were strongly considered.

Transparency, Whistleblowers, and Public Awareness

Witnesses like David Grusch have alleged that key UAP information remains hidden from both Congress and the public. Lawmakers from both parties have pressed for the release of classified materials regarding UAP sightings and any associated government programs. In response, legislation has been proposed to create a central depository for UAP-related records and to establish protocols for whistleblower protections.

❖ Significant Outcomes of the 2024 Congressional Hearings on UAPs

The 2024 congressional hearings on Unidentified Aerial Phenomena represented a pivotal moment in both the governmental efforts for UAP transparency and for the public discourse surrounding unexplained aerial events. Building upon momentum from previous hearings and investigations, the 2024 sessions brought together lawmakers, intelligence officials, military personnel, and expert witnesses to scrutinize the U.S. government's handling of UAPs.

The hearings featured testimonies from high-ranking defense officials, whistleblowers, pilots, and scientists. Significant moments included:

- Firsthand Military Accounts: Several former and active-duty pilots described encounters with objects exhibiting extraordinary flight characteristics. These included sudden accelerations, the rapid onset of hypersonic speed, and the lack of visible means of propulsion, as previously noted.

- Government Whistleblowers: Individuals with inside knowledge testified about alleged government programs investigating UAPs. These included claims of recovered UAP materials and reported efforts to reverse-engineer non-human technology. While these claims stirred controversy, they underscored the need for greater oversight and transparency.

- Intelligence Community Statements: Representatives from the intelligence community acknowledged the existence of dozens of unresolved UAP cases. They emphasized that some incidents remained unexplained even after rigorous analysis. Officials also noted increased efforts to remove the stigma associated with reporting UAP encounters.

- Scientific Perspectives: Scientists called for a systematic, interdisciplinary approach to studying UAPs, emphasizing the importance of open data sharing and collaboration between governmental and civilian research institutions.

Significant Outcomes

1. Expanded Government Transparency

One of the most notable outcomes was a promised commitment to increased governmental transparency regarding UAP investigations. Lawmakers pressed defense officials and intelligence leaders to make more information available to the public. This could be accomplished by the declassification and release of additional classified reports and previously withheld data.

The hearings also led to the drafting of new legislation requiring regular Congressional updates, public briefings, and a streamlined process for pilots and military personnel to report sightings without fear of reprisal.

2. Strengthened Oversight and Accountability

The hearings prompted Congress to strengthen its oversight of the Pentagon's UAP-related programs. Lawmakers authorized an independent review board comprising scientists, military experts, and civilian representatives to evaluate the actions of agencies involved in UAP tracking and analysis. This board is tasked with ensuring compliance with reporting standards, safeguarding whistleblower protections, and investigating any allegations of clandestine programs operating outside official channels.

3. National Security Implications

National security again remained a central concern throughout the hearings. Witnesses and committee members underscored the potential risks posed by UAPs, whether they represent advanced foreign technology, unknown natural phenomena, or something else entirely.

The Department of Defense committed to bolstering sensor networks around sensitive installations and sharing data with allied nations to better monitor and respond to potential threats. The hearings also highlighted the importance of distinguishing between genuine threats and benign anomalies to prevent unnecessary escalation.

4. Encouragement of Scientific Inquiry

A major outcome of the hearings was the call for a more robust scientific approach to UAP investigation. Funding was allocated for interdisciplinary research, and partnerships were established between government agencies, universities, and private sector organizations.

Open calls were made for expanded data sharing and peer-reviewed analysis. This indicated a remarkable shift away from the realms of secrecy and into increased scientific rigor. This collaborative framework is intended to accelerate understanding of UAPs, regardless of their origin.

5. Reduction of Stigma and Cultural Shift

Witnesses and lawmakers alike acknowledged the detrimental effect that has long been held on reporting UAP incidents. This is particularly observed within the military and aviation communities. The hearings catalyzed a cultural shift, making it clear that reporting unusual aerial phenomena is a matter of national interest and public safety, and not a subject of ridicule. Policy changes were needed to protect those who come forward and incentivize accurate, timely reporting.

Since this directly affects what I have now done, I am personally pleased to see this happen.

6. Catalyzing International Cooperation

Recognizing that UAPs are not uniquely an American phenomenon, the 2024 hearings fostered increased international collaboration. The U.S. government extended invitations to foreign intelligence and defense partners to participate in joint investigations and information sharing. Efforts are underway to develop standardized protocols for the collection, analysis, and dissemination of UAP data among allied nations, furthering a unified global approach.

7. Legislative Actions and Funding

As a direct result of the hearings, Congress passed new legislation targeting gaps in UAP oversight, funding sensor technology upgrades, and supporting research initiatives. Appropriations bills included specific earmarks for UAP-related projects, ensuring sustained attention and resources for ongoing investigation.

8. Whistleblower Protections

Several witnesses raised concerns about retaliation against individuals reporting UAP-related information. The hearings have led to the implementation of new whistleblower protection measures, safeguarding those who provide information and promoting an environment where transparency is valued over secrecy. This important debate will persist for the foreseeable future.

The 2024 congressional hearings on UAPs marked yet another watershed moment in government transparency, scientific inquiry, and for national security policy. By fostering open discussion, encouraging rigorous investigation, and prioritizing the protection of those who come forward, lawmakers took significant steps toward addressing the UAP mystery.

While many UAP questions remained unanswered, the outcomes of these hearings laid the foundation for continued exploration, cooperation, and potentially extraordinary new discoveries.

The hearings promised for May 2025 were cancelled, but they have hope for upcoming hearings. This could be important for the possibilities of further progress for the UAP research communities and for curious citizens who continue to follow this issue.

One more of my stories:

The USS Montpelier was a brand-new attack submarine in 1991 and was preparing to go out on sea trials. In order to do so, it docked at the Naval Ordnance Test Unit, known as NOTU. This is at Port Canaveral, owned by the Navy but within the security zone of the Air Force Station. Upon finishing their preparations, they headed out into the Atlantic Ocean to sail down to the Bahamas. While there, they would perform specialized technical procedures that would be required before the sub became fully operational.

Not long after they left port, they wanted to practice a crash dive exercise. Unfortunately, the executive officer took a fall in the process and was injured. The IDMT (Independent Duty Medical Technician) requested a doctor to evaluate this officer, and since I was the flight surgeon of the day, it was my duty to respond.

Through this event, I had the privilege to go aboard the nuclear attack submarine, which was pretty cool. During my time as a visitor to this amazing machine, I had the great opportunity to stand on the sail and watch the dolphins leaping across the bow. I was able to look through the periscope as we submerged, which was also pretty cool.

I got a tour around the sub and even had a short introduction to independently controlling the movements of the $1.7 billion submarine from the inboard helm position. I must say that the car I am driving today is considerably cheaper than the nuclear sub I drove at that time. It is noteworthy that I was still in my blue flight suit with DDMS and space shuttle support team patches that I was wearing that day.

At the end of this adventure, I was put ashore on the island of Eleuthera. I paid an exorbitant fee, as required, to get from the dock to the Rock Sound Airport. From the Bahamas to Miami was a little bit less than 200 miles, so I was pleased to be a passenger on the only aircraft that could get me there that day. The aircraft was a twin-engine propeller plane that had only a single pilot. There were two other passengers that day, and in the heat of the compartment, both of them dozed off fairly quickly after take-off.

Just a few minutes later, I was surprised when the pilot turned toward me over his left shoulder and said, "Hey Doc. Why don't you come up here with me. You can sit here in the left seat and we can talk for a while."

He waved me forward so I slowly made my way up to the left pilot's seat while the actual pilot remained in the right seat. As I sat down, I asked, "How did you know I'm a doctor?"

The pilot pointed to my name tag and explained, "I saw the flight surgeon wings. I used to be a naval aviator so I am familiar with flight surgeons."

I laughed and joked, "Of course a naval aviator is more skilled than any Air Force pilot. I understand your chosen mentality."

He laughed back and stated, "You can't argue with the facts!"

Over the next ten minutes or so, we were chatting about the weather and routine topics. All of a sudden, he asked, "So how long were you on the island?"

I explained, "Oh, I just got here this morning."

He looked at me in an awkward fashion. "But you weren't on my flight when I landed this morning. How did you get on the island?"

Without any considerations, I responded, "Well, I got off a sub."

The pilot appeared more serious now. "I know there were a bunch of scientists and technicians going aboard a sub that was going out for sea trials. Were you on it?"

I smiled and told him, "Yes. I got on the Montpelier as it was coming out here after it left from Port Canaveral, at NOTU. As they were leaving port, they tried to have an emergency crash dive exercise but had the XO slip and get hurt. He was okay, but that provided me the opportunity to go aboard the sub, providing that I would have to pay for my flight back to Florida."

He appeared incredulous as he repeated, "You got on the sub because the XO got hurt?"

My response was to say, "They had a brand new IDMT and when he saw that the XO was hurt, he requested that a doctor needed to check him out. Since they had just left NOTU, the only doctor available was me. I was the 45th Medical Group's flight surgeon of the day so I was loaned to the Navy for that reason. The rest of everything was just gravy for me. I had always wanted to be on a sub, so when I got this chance, I took it."

Studying the patches on my flight suit, he inquired, "But you work with the space shuttle? I see what you are wearing, so how does that work?"

I then explained to him, "I am the Chief of Aerospace Medicine for the Eastern Space and Missile Center, located at Patrick AFB. In doing this, I support space shuttle launch and landing operations for NASA, through the Department of Defense Manned Spaceflight Support Office, called DDMS. But I also performed SAR missions with the 41st Air Rescue

Squadron. I likewise supported AFTAC, the State Department mission based out of Patrick, the U-2 operations from the 9th Strategic Reconnaissance Wing, the Joint STARS developmental program, and every other function we manage. We support NOTU, but this is the only time I have had the chance to be on a submarine."

Pausing for consideration, the aviator wondered, "So because the XO got hurt on this sub, you went aboard and came all the way out here to Eleuthera?"

I shrugged and told him, "That's what happened. It was just a quirk of events or I still would have been in my flight medicine clinic at Patrick."

He waited, then lowered his voice. "Look, Doc, I did quite a few classified missions while I was in the Navy, so you can tell me what really happened. I understand about your cover story, so if anyone would ask, I will tell them what you said."

My brow furled and I replied, "No. It's nothing like that. It is just that the officer was injured, and they wanted me to check him out. There is nothing else going on."

This former naval aviator was not convinced. "Trust me, Doc, I can keep my mouth shut. Let me know what was really going on!"

For several minutes, we went back and forth on this. Finally, he promised me, "Look, Doc, we've got a straight shot to Miami. If you tell me what you were really doing, I'll let you fly us back to the airport. At least, until we enter the traffic pattern. This is not a complicated aircraft to fly."

After repeating the truth several more times but being faced with a total lack of his belief in the story, I finally agreed, "Show me the controls and I'll fly us back in."

He eagerly went through the controls with me and, as he described, it was a simple control setup compared to our military aircraft. Once he positively transferred the controls to me, he said, "So what were you really doing?"

Because the truth had failed miserably, I announced to him, "The Air Force has a secret airbase in the Bahamas, which you know is in the Bermuda Triangle. Well, NASA assists the Air Force in monitoring UFO bases down here. It is really 'hush, hush' as you would imagine."

The pleased pilot clapped his hands and proclaimed, "I <u>knew</u> that was going on down there! I was told that all of those scientists and technicians were needed just for the submarine operations, but I knew there was more to the story!"

I eventually returned the controls to the happy fellow and he landed us uneventfully. He then shook my hands vigorously after landing as we parted ways on the parking tarmac for his small airline.

I rented a car from the airport and drove back up the coast to get back to Patrick Air Force Base and to my own home. Since then, this is a story I have told to my family and a few close friends as the years passed.

What amazed me the most was that he absolutely refused to believe the truth. When I told him a completely fabricated tale, he was as delighted as he could be.

However, at that time, I had never seen a flying saucer or any other form of UFO. Little did I know that about a year later, there would be a subsequent event on Cape Canaveral Air Force Station in which I would truly see the very type of spacecraft I had teasingly just disclosed!

Who knew?

We Got it From Them!

Chapter 4

The Science of Observing UFOs

Science relies on verifiable evidence and repeatable observations. UFO reports challenge these principles, as they are often anecdotal, fleeting, and difficult to document. Nevertheless, scientists can still approach UFO cases through several disciplines:

- Physics: Calculations of observed speed, trajectory, and maneuverability of UFOs often conflict with currently known laws of aerodynamics.

- Astronomy: Many UFO sightings are ultimately identified as being astronomical objects like planets, meteors, satellites, or atmospheric phenomena.

- Psychology: Cognitive biases, perceptual errors, and social influences might play a significant role in misidentification and mass sightings. If a person believes in UFOs and eagerly seeks to find a UFO, this bias could affect their perception. They may tend to interpret common events as UFO sightings.

- Radar and Sensor Analysis: Some UFOs are tracked by radar, lending credibility to certain reports, but sensor glitches and false positives are common. It is the quality of the radar systems and their operators that determine the validity value of the findings.

The Role of Technology

Smartphones, dashcams, and advanced telescopes have made it easier than ever to record unusual aerial events. However, increased documentation has also led to a deluge of hoaxes, misinterpretations, and altered footage, complicating scientific analysis.

❖ NASA's Reporting and Study of Unidentified Anomalous Phenomena

Let us explore NASA's evaluations and capabilities for their new interest in UAPs. While we have discussed this before, I want to go deeper into what they could do, if they are truly interested in this subject.

This recent shift that was thrust upon them is a fundamental change in NASA's approach to being open about identifying UAPs. This has not been their prevailing attitude in years gone by. This reflects a broader intent to capture not only the unexplained aerial objects, but also any strange phenomena that may been noted within the air, sea, space, and other environments. Increased public interest has been spurred by declassified U.S. military footage, official government reports, and congressional hearings.

Now that it has also affected NASA, I will be most interested to watch and hear how they change their culture. They have not always been eager to acknowledge that there are real phenomena observed in our skies. Because of this, it will be quite interesting to see how they will assess their own stated lack of suitable conventional explanations.

Formation of the UAP Independent Study Team

In 2022, NASA announced the formation of an independent study team dedicated to investigating UAPs. The team, composed of leading scientists, data analysts, and aviation safety experts, was tasked with evaluating existing UAP data. This would identify the best ways NASA could contribute to further understanding, and to developing a roadmap for advancing the scientific study of these occurrences. This ongoing effort was designed to complement other federal initiatives, and to promote transparency and rigorous investigation within NASA itself.

Objectives of the Study

- To assemble and analyze all publicly available UAP data.

- To assess data collection methodologies best suited for future UAP studies.

- To recommend a roadmap for NASA's potential role in joint UAP research projects with other governmental agencies.

- To foster transparent public communication about any new UAP findings and methods.

Key Aspects of NASA's Reporting on UAPs

1. Scientific Methodology

NASA's approach to UAPs is to be heavily centered on scientific objectivity. The agency's reporting emphasizes that, so far, there is no credible evidence to suggest that UAPs are of extraterrestrial origin. It is important to understand this institutional bias, for it has been deeply embedded into their entire culture for decades.

Instead, NASA professes to support a newer, hypothesis-driven approach. They plan to increase their ability to collect robust data, eliminate routine explanations, and only then consider the more extraordinary possibilities for their cases. This supposed discipline is meant to ensure that UAP studies avoid sensationalism and speculation.

2. Data Collection and Challenges

One of NASA's central findings is that most existing UAP data suffer from poor quality, lack of metadata, and inconsistent reporting standards. Many UAP reports come from military and civilian pilots, law enforcement, or the general public, but are often lacking precise instrumentation or objective measurements. NASA's reporting underscores the need for:

- High-quality, calibrated data from multiple sources.

- Standardized reporting forms and protocols.

- Advanced analysis tools, including machine learning and artificial intelligence, to identify hidden patterns and rule out known causes.

3. Collaboration and Transparency

NASA's reporting process is characterized by its proposed commitment to transparency. Unlike previous eras, when information about UAPs was often classified or shrouded in secrecy, NASA has stated they will now openly publish their findings, methodologies, and recommendations. The independent study team's final report, released to the public, details not only its conclusions but also its challenges and limitations.

NASA also collaborates with other federal agencies, including the Department of Defense's All-domain Anomaly Resolution Office, to share data and analytical techniques.

My key concern here is that NASA and AARO may choose to follow the same approaches and use the same mentality in their analyses. If this occurs, we will simply have two more agencies following the old-time governmental idea of reporting what they want us to believe, rather than what the raw data truly indicates.

4. Public Engagement

Recognizing the high level of public fascination, NASA has prioritized clear and accessible communication. The agency has held public meetings, press briefings, and Q&A sessions to demystify its work on UAPs. NASA's reporting emphasizes that science thrives on open questions and public discourse, inviting citizens to participate by reporting UAP sightings through appropriate channels.

Key Findings from NASA's UAP Reporting

- No Evidence of Extraterrestrial Origin: To date, NASA has found no evidence in their studies and analyses to support the idea that UAPs are linked to alien technology or non-human life. This is highly consistent to what other governmental organizations have always reported. This is no surprise!

- Insufficient Data Quality: The majority of UAP cases remain unexplained. NASA relates this to poor quality or incomplete data, rather than any inherently mysterious properties.

- Potential Explanations: Many UAPs are eventually attributed to the same explanations used by the Air Force's Project Sign, Project Grudge, and Project Blue Book. These would be airborne clutter, commercial aircraft, drones, weather phenomena, optical illusions, or sensor anomalies. However, some cases remain unresolved. It is these unresolved cases that UAP enthusiasts identify as the reason NASA should follow up with increased levels of scrutiny.

This is an example of the source for so much frustration within the UAP communities. When we see and hear the government agencies state the same old tired phrases, and supply the same old preset conclusions, it certainly reminds me of all the hidden agendas that have served them for nearly eighty years. This is why we have so little faith in the system.

If they truly don't know what any specific anomaly is, they should dig deeper into that study. Instead, what these official reports tend to do, over a period of decades, is to simply say "We don't know!" Yet they <u>always</u> choose to report that there is no evidence of the extraterrestrial nature of UAPs.

I must ask, if you don't really know what a mysterious craft is, how can you truthfully conclude that it is absolutely <u>not</u> extraterrestrial?

NASA, be better than that!

Recommendations for Future Investigations: NASA advocates for improved sensor calibration, standardized data collection, and the application of novel analytical tools to enhance the scientific rigor of future reporting. To do this, NASA should follow their own proposals for using advanced technology to enhance the serious study of UAPs:

- Develop and Deploy Specialized Sensors: Invest in the development of sensors specifically designed to observe UAPs across multiple wavelengths and domains (air, sea, space).

- Establish Standardized Reporting Protocols: Create unified forms and databases accessible to both governmental agencies and the public to ensure consistency and comparability in UAP reports.

- Promote Multidisciplinary Research: Encourage collaboration between universities and governmental agencies to provide coordination between legitimate experts in fields such as meteorology, aerospace engineering, astrophysics, and artificial intelligence.

- Ensure Transparency and Public Access: Make their data and findings widely available, subject to privacy and national security considerations. In doing this, they could foster trust and enable independent verification, regardless of the findings.

- Engage with the International Community: Work with space agencies and researchers worldwide to share data, harmonize standards, and collectively address UAPs as a global scientific challenge. It is necessary that all governments share their UAP data equally!

NASA's recent involvement in reporting and investigating UAPs marks a truly significant moment in the history of their scientific inquiry into unexplained phenomena. By applying rigorous scientific methods, promoting transparency, and inviting public participation, NASA has transformed their historical culture's conversation around UAPs. They need to shift their focus from one of secrecy and hidden speculation to one of unhindered inquiry and evidence-based analysis that will be open to the public.

At least, this is surely what we will hope for from the space agency!

Yet Another Personal Story:

I have mentioned the Air Force Technical Application Center, AFTAC, earlier in this book. I will not be discussing the depth of their mission or operations except to say that I had deployed on a mission with them flying in an Air Force WC-135 "Weather plane."

As part of our course of flight, we landed in the South Atlantic on Ascension Island. We were only able to do this because we had permission from the British government. There was a good reason for this, so allow me to provide some information.

Ascension Island is a UK volcanic military outpost nestled far out within the South Atlantic Ocean. It is a very unique and captivating location. For one thing, it is primarily made of reddish rock that makes it look very much like the face of Mars!

Ascension is part of the British Overseas Territory of Saint Helena and Tristan da Cunha. This odd island boasts a remarkable history, an extraordinary ecosystem, and remains of critical importance for military stability of the Southern Atlantic region. It also plays a strategic role within global communications and scientific research, as I will explain below.

It is situated approximately 1,000 miles (or 1,600 kilometers) from the western coast of Africa and roughly 1,400 miles (2,250 kilometers) from the eastern coast of Brazil. Covering an area of just over 34 square miles (88 square kilometers), it is a small yet geographically diverse island. Its volcanic origins are evident in its rugged terrain, with craters, lava flows, and barren landscapes juxtaposed with lush, man-made greenery. The island's highest peak, Green Mountain, rises to 2,818 feet (859 meters) and is a striking feature amid the arid environment. For one thing, it is the only spot I saw on the island that had any major area of green vegetation.

The history of Ascension Island is as intriguing as its geography. The island was first discovered by the Portuguese explorer João da Nova in 1501. Despite this, it actually received its name from a totally different Portuguese explorer. Afonso de Albuquerque found it in 1503 and arrived there on the Christian holiday of Ascension Day. He then named the island because of this event. For centuries, the island remained uninhabited, serving only as a stopover for ships navigating the southern Atlantic Ocean.

In 1815, Ascension Island gained strategic importance when the British established a garrison to prevent any rescue attempts of Napoleon Bonaparte, who was exiled to Saint Helena, about 800 miles (or nearly 1,300 kilometers) away. Over time, the island transitioned from a military outpost to a key hub for global communications and research.

There are no native inhabitants from the island and its only residents are those belonging to the UK military who are assigned there. The island continues to host a variety of facilities, which supports military and civilian operations, along with satellite tracking stations used by space agencies such as NASA and the European Space Agency. The island also serves as a base for scientific research, including studies on climate change, marine ecology, and volcanic activity.

The island is home to unique fauna, including the Ascension frigatebird, which nests exclusively on the island. There are green sea turtles which migrate thousands of miles to lay eggs on its beaches. Efforts to protect these species have been bolstered in recent years, with the Ascension Island Marine Protected Area being one of the largest marine conservation zones in the world. It was established in 2019. While not native to the island, sheep herds were present when our aircraft landed at their Wideawake Airfield.

This brings me to the crux of why I am repeating this story. Our aircraft landed at the airfield just before dark. We took our gear to some very simple concrete-brick barracks. There was, sadly, no air conditioning in the place. After having the bright sun shining on them all day, we opened all of the doors and windows in the barracks to hopefully allow the cooler night air into the rooms in which we would be sleeping.

My aircrew were given the best hospitality possible, considering how late in the day we had landed. This primarily consisted of allowing us into the officers' club, where sandwiches and soft drinks were provided. The main source of entertainment was a pair of billiard tables. Rather than playing traditional pool games, everyone began to play a military-style game called crud.

The rules for crud were simple. To start, the two opponents begin from opposite ends of the pool table, with one holding a single solid colored ball while the other held a single striped ball. In the center of the table, the solid white cue ball was placed. Once the game had begun, the two opponents would attempt to knock the cue ball into one of the two pockets that belonged to their challenger. You would strike the cue ball with the solid or striped ball you held, with the use of force fully authorized. This was not a gentleman's game. It was a poorly controlled combat contest. Bumps, bloody noses, and even abrasions and lacerations were frequently experienced.

Tempers were easily set off and the games became ever more violent over about thirty minutes or so. At that point, the senior officers for both the British and American personnel had to put a stop on this game. It seemed to me that we were on the verge of having an international incident as the warring game-players were ready to break out into combat assaults against each other.

Following this, we Americans headed for our assigned barracks. The rooms were still very warm so we stripped down to our underwear and tried to lay down for any amount of sleep we might be able to attain that evening. I could hear griping from the various rooms down the hallway since the temperatures were not at all pleasant. I would have complained, but it would have made no difference in our circumstances.

At about two hours later, I suddenly heard a very loud banging sound coming down the hall from the end of the hallway. All kinds of strange sounds were breaking out, so I turned on my room's light and went out into the hallway. To my complete surprise, I was greeted by a host of large sheep that were running down the hallway from my left toward the right. I quickly stepped back into my room but two sheep followed me into this small space before I could close the door. I jumped up onto my bed and grabbed my pillow and started swinging it at these invaders, and they finally left. Before I could close my door, the previous sheep were replaced by yet another one. I continued to defend my room as everyone else fought their battles wherever they found themselves. After about ten minutes, we cleared the last of the animals out the door to the right end of our barracks.

We made the collective decision to close the doors at each end of the hallway, regardless of the heat and decreased airflow that would have possibly assisted us in cooling down our sleeping quarters. I must say, when we all climbed into our aircraft the next morning, between the wounds from playing crud and the miserable sleeping arrangements we had received due to the sheep invasion and heat, our crew looked pretty poor. As the flight surgeon, I spoke to the aircrew commander and we discussed the complete lack of adequate crew rest, yet we both agreed that it was best to get on our weather plane and head toward South America, to complete our mission.

There were no purple hearts given out that morning, but it truly felt as though we had completed a combat deployment!

We Got it From Them!

Chapter 5

Natural and Man-Made Explanations Against UAP Theories

Most UFO sightings eventually yield rather common explanations. This conclusion has been confirmed by a multitude of governmental sources and even civilian UFO research organizations. With this in mind, the following categories account for a large proportion of these cases:

- Astronomical Bodies: Venus, Mars, and persistently bright stars are frequently mistaken for UFOs, especially under unusual atmospheric conditions. Also, when the witnesses are moving in vehicles, the motion of cars, trucks, trains, and planes can all act to lead the people to misinterpretations of false or misunderstood motion for what they see. One such example is the well-known effect of parallax motion.

- Atmospheric Phenomena: Ball lightning, auroras, and lenticular clouds can create startling visual effects. To this I can add my own experience. During one of my military flights, two of the pilots in our formation briefly thought they were seeing an actual UFO in the distance. As we studied the object more carefully and approached it closer, we were able to determine that it was only a lenticular cloud. For the first few seconds, however, that cloud fooled us.

- Aircraft and Drones: Military exercises, experimental aircraft, and civilian drones are often reported as UFOs, particularly when seen at a distance or in unfamiliar lighting. With the wide range of drones available today, including such varieties of small and rather large ones now flying, they can be mistaken for both manned aircraft and possible UFOs.

- Satellites and Space Debris: The International Space Station, Starlink satellites, and re-entering debris have generated waves of UFO reports around the globe. At dawn and dusk, when people are in relative darkness on the ground, brightly lit satellites can become more obvious to observers. We will have some space debris re-entering the Earth's atmosphere every single day. But under certain conditions, we may catch a glimpse of their unique appearance as they do so.

- Human Artifacts: Weather balloons, flares, and kites are sometimes misperceived as mysterious flying objects. Particularly when you consider the range of human imaginations, there can be balloons and kites with all types of extraordinary configurations. Even Chinese lanterns, powered only by the heat from a candle, can sometimes be seen at night as floating lights or orbs.

Examining the Skeptical Perspective on Unidentified Flying Objects

It is time to explore the primary lines of rebuttal evidence that so many scientists and skeptics present in arguing against common UFO beliefs. We will see how they dissect eyewitness accounts, along with much of the photographic and video evidence, the radar data, the physical traces to be found, and especially unexplained artifacts.

It is a foundational principle in science that extraordinary claims require extraordinary evidence. The assertion that UFOs are spacecraft from other worlds represents a profound challenge to our current understanding of physics, astronomy, and biology. As such, scientists demand not only credible evidence, but also evidence that is robust, reproducible, and withstands critical scrutiny. This ethos underpins the scientific rebuttal to most UFO claims.

Eyewitness Testimony: Unreliable and Susceptible to Error

Eyewitnesses are often at the heart of UFO reports, providing vivid and sometimes dramatic accounts of mysterious aerial phenomena. Even in legal matters such as bank robberies and murder cases, eyewitness reports often prove unreliable. Scientists point to the several decades of their psychological research that has been performed that has demonstrated how human perception and memory, especially from years past, are ofttimes inherently flawed.

- Perceptual Errors: Factors such as atmospheric conditions, lighting, distance, and the speed of the observed object can lead to misidentification of commonplace objects as being of some extraordinary design.

- Memory Distortion: Memories are not static recordings but are reconstructed over time. Stress, the passage of time, and social influences can all alter or embellish an original experience. It has been shown that by simply retelling the same story over and over again, especially when this occurs through a number of years, this can lead to new discrepancies compared to their original reports. This is not intentional, but it is how our brains work.

- Influences of Expectation and Suggestion: Cultural narratives, media coverage, and group dynamics can significantly shape how individuals interpret ambiguous stimuli. Studies show that people are more likely to interpret a light in the sky as a UFO if they already harbor beliefs in extraterrestrial visitations.

Because of these factors, scientists generally regard anecdotal reports as the weakest form of evidence, especially in the absence of corroborating physical data. Unfortunately, with a transient UFO encounter, the event that happens is often experienced so rapidly that the mind may be unable to process all of the details correctly. Numerous police and legal television shows and movies have demonstrated this to us.

Photographic and Video Evidence: The Limits of Visual Documentation

Photographs and videos are commonly presented as "proof" of UFOs. Yet, such evidence is fraught with the following problems:

- Quality and Ambiguity: The vast majority of UFO images are blurry, distant, and taken in poor lighting conditions. They commonly involve inferior cameras, often simply cell phones, which makes definitive identification and thorough image processing nearly impossible.

- Technological Manipulation: With advances in photo and video editing software, it is easier than ever to intentionally create convincing fabrications. Even unintentional changes can misrepresent ordinary objects, such as nearby bugs, flocks of birds, unusual drones, or aircraft that are seen in unusual lighting environments. They will seem to be extraordinary when truthfully they certainly are not.

- Lack of Scale and Context: A photograph or video, particularly when taken hurriedly, often lacks reference points or similar identifying information. This can make it extremely difficult to properly judge size, speed, and distance. As a result, commonplace things like weather balloons, aircraft lights, or even insects very close to the lens, can appear mysterious.

- Confirmed Hoaxes: A history of hoaxing in the UFO community undermines the reliability of photographic evidence, as even a few false cases taint the credibility of the corpus as a whole. Since the earliest days of the UFO phenomenon, there have been unscrupulous individuals who have gotten their kicks in this fashion. A few have even made a fair amount of money by doing so. These people make it extremely difficult for serious researchers to avoid the damage done by tricksters and sometimes criminals.

Radar and Instrumental Data: Conventional Explanations Abound

Some UFO proponents point to abnormal radar returns or unexplained blips as evidence of anomalous craft. Skeptical scientists offer several natural and technological explanations for most radar anomalies. What are some things that can be responsible for radar abnormalities?

Radar reflections and atmospheric phenomena can be the cause of false returns. Temperature inversions, weather fronts, and even flocks of birds can produce radar signatures that seem abnormal but are entirely natural. Yet as radar systems become more advanced and radar controllers become more proficient at reading the returns on their screens (especially as they are integrating AI into the analysis picture), we can hope that misinterpretations will diminish. If they do, we will have more verifiable data available to us for deeper studies.

Without physical evidence that links radar blips to tangible craft, skeptical scientists regard such data as completely insufficient to justify their belief in extraterrestrial visitations. That is just the way it is with scientific research. History is filled with well-meaning scientists who have promoted their new ideas, only to be shot down by later discoveries.

Physical Traces and Artifacts: The Elusive Evidence

Occasionally, reports emerge claiming that UFOs have left behind physical traces such as burned vegetation, circular impressions, signs of radioactive exposures, or even "alien" artifacts. However, these claims face significant logical challenges from the scientific community.

- The Lack of Unique Signatures: Physical traces that have been initially attributed to UFOs have, upon investigation, sometimes been explained as resulting from routine human activity, such as vehicle tracks or past campfires. Even animal behavior or meteorological phenomena have been documented to create abnormal findings that do not meet the criteria for UFO experiences.

- Absence of Material Evidence: No artifact or material sample presented as "alien" has ever withstood rigorous scientific testing. Claims of recovered alien technology have consistently failed to produce verifiable, replicable results. How much of this "lack of evidence" is due to the government's intentional cover-up is a matter of strong debate.

Take the Kecksburg event as such an example. There may have been no residual evidence because military personnel had already recovered the vehicle. Hauling away the craft, and anything related to it, on the reported flatbed truck would remove the very evidence that would be required by later investigations.

This could be exactly why there was no residual evidence that was left at the scene. When such evidence is recovered and hidden within clandestine governmental vaults, there will be no evidence remaining to show to the general public.

- Problems with Proper Chain of Custody Elements and Accidental Contamination: Many reported artifacts are not preserved utilizing proper scientific and legal protocols, raising the possibility of contamination or tampering. Whether in an actual court or in the court of public opinion, ensuring the accuracy of data and recovered materials is all important.

Despite decades of reports, the scientific community continues to state they have never received credible, peer-reviewed evidence of extraterrestrial materials. These types of statements are not generally received with enthusiasm by the UFO communities. The trust levels that many people hold in our governmental agencies today are at an all-time low.

In the end, your personal beliefs will largely shape how you view the supposed evidence. However, missing evidence that has been sequestered away by government agencies is not at all the same as evidence that never existed. Which end of this spectrum will you choose to believe? Which evidence, or lack thereof, will you accept as truth?

Therein lies the rub!

My Adventure During a Winter REFORGER

In 1986, we held a Winter REFORGER (Return of Forces to Germany) exercise that I remember well, for several reasons. These massive REFORGER exercises were a series of large-scale military maneuvers held annually during the Cold War, primarily in West Germany. These exercises were organized by NATO (North Atlantic Treaty Organization) with a focus on rapidly deploying sufficient American troops to Europe in the event of a conflict with the Soviet Union. At that time, I was stationed in West Germany and World War III seemed nearly inevitable.

Considering the recent invasion of the Ukraine by Russia in our times, we should understand the necessity for these plans. They served as a clear demonstration of the allied nations' readiness, logistical efficiency, and combined military strength in one of the most worrisome geopolitical regions of the last half of the 20th century.

These massive exercises were specifically meant to send a firm message to the Soviet bloc. Adding to the general tensions of those days, I also supported missions transferring in the Pershing II Intermediate Range Ballistic Missiles (IRBMs). These nuclear IRBM missiles caused so much furor by this action that the U.S. and Soviets eventually destroyed this entire class of nuclear missiles in 1991.

The Purpose of Winter REFORGER

The primary objective of any REFORGER exercise was to reinforce NATO's military presence in Europe and to ensure that the alliance could respond quickly to any external aggression. These operations assessed the ability of U.S. forces to deploy sufficient numbers of troops and equipment to West Germany within a compressed timeframe. This was important in affirming America's commitment to European security. The exercises also sought to reassure NATO allies of U.S. support, while deterring any potential Soviet aggression by showcasing our superior readiness and coordination.

The Historical Context

The REFORGER exercises were initiated in 1969, a year after the Soviet-led Warsaw Pact invasion of Czechoslovakia was witnessed during the Prague Spring. That Soviet aggression heightened fears of further Soviet expansions into Western Europe. These maneuvers were a response to the perceived need for NATO to quickly bolster its defensive capabilities.

The Soviets were believed to have an advantage in the number of tanks over those own by NATO. This was a key to our war plans during those days. Whether it was a summer or winter REFORGER, these exercises became an essential part of NATO's strategy of deterrence. During the later stages of the Cold War, when I was there, the dreadful specter of a potential conflict between the East and West loomed large.

Scale and Scope

The winter REFORGER exercises were among the largest peacetime military operations conducted in Europe. They involved tens of thousands of troops, hundreds of military vehicles, and complex logistical preparations. Both active duty and reserve units were mobilized from widespread bases in the United States, transported all the way across the Atlantic, and then integrated into the standing NATO forces residing in West Germany. These operations emphasized combined arms warfare, with the participation of infantry, armor, artillery, and air support, as well as the coordination of multinational forces. The worst-case scenario would involve a nuclear exchange, and this was not desired by those leaders in the East or the West.

Logistical Challenges

One of the critical elements of the Winter REFORGER exercises was the logistical complexity involved in rapidly deploying and sustaining large numbers of American military forces in a foreign theater. This required meticulous planning, seamless coordination between different branches of the military, and effective use of transportation infrastructure, such as ports, railways, and airfields.

Realistic Training Scenarios

The exercises featured highly realistic scenarios designed to simulate potential conflicts with Warsaw Pact forces. These included defensive and offensive operations, urban warfare training, and joint maneuvers among all of the NATO allies. Soldiers were often required to endure harsh winter conditions, which added an additional layer of challenge and authenticity to the drills.

Integration with NATO Forces

These REFORGER exercises were not solely U.S. operations. They involved extensive collaboration with other NATO member states. This included joint training exercises, the sharing of intelligence, and the development of interoperable communication systems. The exercises helped to strengthen the multi-national cohesion and trust among our NATO allies.

Boosting NATO's Credibility

The REFORGER exercises played a crucial role in reinforcing NATO's credibility as a military alliance. They demonstrated the alliance's ability to mobilize and sustain a large fighting force in Europe, thereby deterring potential aggression from the Soviet Union and its allies. On the other side, the Warsaw Pact countries held similar exercises, aimed to warn us of their own ability to fight us, no matter the conditions.

For the U.S. and NATO forces, these exercises provided invaluable training opportunities, allowing them to hone their skills, test new equipment, and identify areas for improvement. The experience gained from these operations proved instrumental in improving the overall readiness and effectiveness of allied forces. Short of actual war, these were the most realistic exercises we could muster for training and internal cohesion.

Psychological and Political Dimensions

Beyond their military objectives, these REFORGER exercises had significant psychological and political implications. This was especially true to the Germans, who lived in the territories that would be home to these massive battles. They served as a visible and powerful reminder of NATO's unity and resolve, sending a clear message to both allies and adversaries. Additionally, they demonstrated the importance of transatlantic cooperation in maintaining European security.

Challenges and Criticisms

Despite their success, the REFORGER exercises were not without their challenges and serious criticisms. The large-scale movements of troops and equipment often caused disruptions in civilian life, particularly in rural areas of West Germany. There were also concerns about the environmental impact of the maneuvers, as well as the considerable financial costs associated with maintaining such a high level of readiness.

Add in the fact that the middle of the 1980s also included the deployment of the Pershing II Intermediate Range Ballistic Missiles, as I stated before, that could carry nuclear warheads for a range of 1,056 miles (or 1,700 kilometers). The people of West Germany were particularly wary of a war being fought on their own territory. Some protesters even chanted that the shorter the range of nuclear weapons, the deader would be the Germans!

After the fall of the Berlin Wall in 1989 and the subsequent dissolution of the Soviet Union, the recognized need for large-scale exercises, like REFORGER, diminished. The last exercise was held in 1993, marking the end of an era in NATO's military history. However, the lessons learned from these operations continue to influence modern military planning and strategy.

With all of this as a backdrop, now I will add my recollections of this year's events.

The weather was not cooperative for us in 1986, but it was a winter in Central Europe so that was not unexpected. My unit was to be repositioned to the area just north of the infamous Fulda Gap. During the Cold War, the Fulda Gap was considered to be one of the most likely avenues for a Soviet ground invasion of Western Europe. Soviet forces stationed in East Germany could use this gap to advance through this region more easily.

We expected them break through this breach and assault Bonn to the west, Frankfurt to the south, and the heart of the Rhine River Valley as their primary objective. This expected thrust of Soviet forces risked them cutting a key swath through the major economic and logistical hubs of West Germany.

NATO planners saw the Fulda Gap as a key defensive chokepoint. If Soviet forces breached this region, they could potentially outflank

NATO's defenses and disrupt supply lines, posing a serious threat to all of Western Europe. Consequently, holding control of the Fulda Gap became a centerpiece of NATO's defensive strategy, with significant resources allocated to fortifying the area. This was the very reason our wargames that year focused on this important region.

Being part of the Third Infantry Division, our air assets were to be joined along with our ground components at the start of the exercise. However, the weather was such that our first two attempts to fly there were thwarted by the wintery conditions. On our third attempt, we were determined to join with the rest of our troops.

I was flying in the lead aircraft of five OH-58 Kiowa scout helicopters, planned for a V-formation flight. At the time of take-off, we barely had visual minimums to be in the air. As we made our way northwest, the snow got thicker, and our formation had difficulties in keeping the helicopters in view. I will strive to recall the details as factually as I can.

Suddenly, our visual capabilities totally crumped. None of our pilots could see where the other aircraft were, so the flight commander, in my helicopter, ordered a dispersal, as we had briefed before take-off. As Charlie One, we continued straight ahead. Charlie Two, to our left, was supposed to break away with a thirty-degree left bank for thirty seconds. Charlie Three did the same to our right side. Charlie Four was directed to break into a sixty-degree left bank for one minute, then resume our scheduled heading. Charlie Five was to do this, to the right side. This pattern would theoretically place distance between all the aircraft at a safe separation to prevent a mid-air collision.

I was at the controls of our aircraft as the commander tried to herd all of our fellow aircrews ahead. We were all hoping and praying that the weather would clear somewhat. If that happened, the outer aircraft could fly back toward us to reform into a normal flight pattern. This was becoming a much more difficult task by the second as our vision became worse by the minute.

I was in the right seat and just happened to catch a dark blur out of my right lateral vision. In just a split second, I was alarmed to see that the Charlie Three aircraft was right out my door and was crossing our path!

I immediately lifted the collective to increase the lift to our aircraft, adjusting for this increased power by pressing the left pedal, which altered the force of our tail rotor. I pulled way back on the cyclic, causing our rotor disk to assume a high angle-of-attack, which swiftly increased our altitude. Just then, the other scout helicopter passed right in front of our feet, seemingly only about thirty or forty feet from us!

I tersely called out, "Traffic!"

The flight commander in the left seat was as wide-eyed as I imagined I was as well. As the other helicopter quickly disappeared from our sight, the commander called out over the radio to our flight, "Vertical dispersal!"

When our helicopters had previously dispersed at thirty-degree increments, we were all maintaining the same elevation. With this order, our pre-flight briefing went into effect. We, as Charlie One, would maintain our heading and altitude. Charlies Two and Three would climb by three hundred feet to give us vertical separation. Charlies Four and Five were to rise by five hundred feet, giving them additional vertical separation from Two and Three.

Moments later, the flight commander commented to me that because Charlie Three had come from our right, he had not seen the helicopter at all. He shook his head and radioed, "One will RTB (return to base) immediately. Two and Three will follow in thirty seconds and Four and Five will follow in one minute. Everyone, fly with caution!"

We were in an area where the highest flight obstacle was about 500 feet, so we dropped down to 600 feet above our suspected ground level. Finally, we could partially see the ground, so we proceeded back to our airfield with no further major events. Eventually, visibility improved near our home airfield and all five aircraft returned safely. In the debrief, the crew of Charlie Three reported that they had never seen our aircraft after dispersal, so they had no idea how close both of our aircrews came to dying within those snow clouds.

Even in peacetime, tactical flying within the military is dangerous, for mishaps can occur just that quickly!

Chapter 6

The "Unexplained" Cases

A persistent mystery lingers over a small subset of known UFO cases that were first identified by Project Blue Book. These cases resist any type of straightforward explanations, and we continue to find such cases happening even today.

These outlying cases included inexplicable incidents that often involve multiple witnesses, simultaneous radar and visual confirmations, and those leaving behind detectible physical effects. These include scorched patches of earth, noted physical changes to trees or landscapes, or even electromagnetic interference effects to nearby electronic equipment.

Military Encounters

In recent years, the U.S. Department of Defense has confirmed several Navy pilot encounters with UAPs that exhibited extraordinary speed and maneuverability. As we have reviewed, declassified videos clearly show these objects accelerating, stopping, and changing directions in ways that defy conventional engineering and flight principles. These persisting incidents have prompted renewed interest in the scientific investigations at the highest levels of our government.

Physical Evidence

While most UFO encounters leave no trace, a handful of cases involve unexplained physical phenomena. These may include lingering metallic debris, anomalous radiation levels, or visible changes to soil and vegetation. Scientific analysis of these materials has, so far, not produced conclusive, accepted, evidence of exotic technology.

❖ What is Our Best Evidence of Real UFOs/UAPs?

For more than seven decades, UFOs and UAPs have captivated both the public imagination and the serious scrutiny of scientists, governments, and military personnel. While many sightings can be explained by natural phenomena, human error, or man-made objects, a small subset of cases remain quite baffling despite rigorous scientific investigations. These cases, which include credible witness testimonies, radar sensor data, and photographic or video evidence, form the core of what many consider the "best evidence" for real UAPs.

Let us present a few of the detailed explorations of the most credible and enduring evidence for real UAPs, summarizing historic and recent cases. This will include declassified government reports, military encounters, and independent scientific analyses.

1. Historical Context of UAP/UFO Encounters

We know that the modern era of UFO sightings began in 1947. This was when pilot Kenneth Arnold reported seeing nine disc-shaped objects over Mount Rainier, Washington. Arnold's reported description led to the popularization of the term "flying saucer." In the years that followed, a wave of sightings, both in the United States and internationally, spurred government investigations such as Project Blue Book and attracted significant media attention.

Project Blue Book & Government Studies

The United States Air Force initiated Project Sign in 1947, then followed on by introducing Project Grudge in 1949. Then, in March of 1952, Project Blue Book was created to "scientifically" study all these incoming UFO reports. The new project was originally headquartered at the Wright-Patterson Air Force Base in Ohio. It was initially led by Captain Edward J. Ruppelt, but it is best known due to the presence of the renowned astronomer J. Allen Hynek.

Over its 17-year run, Blue Book collected more than 12,000 sightings. While the majority were explained as natural or manmade in nature, approximately 701 cases, about 6%, remained officially "unidentified." Many skeptics pointed out the lack of physical evidence, yet the persistence of so many unknown cases in such a large dataset remained intriguing.

Early on, Hynek was used to debunk many of these cases. However, as time passed and the reports continued, Hynek became known as one of the most authoritative UFO advocates during his lifetime. He even played a small role in the movie "Close Encounters of the Third Kind."

2. Military Encounters and Sensor Data

The Breakthrough 2004 USS Nimitz Encounter, Known as the "Tic Tac" UAP Event

One of the most compelling modern cases involves the 2004 encounter between the USS Nimitz Carrier Strike Group and a series of unidentified aerial objects detected flying off the coast of Southern California. We have already mentioned that multiple highly trained Navy pilots, including Commander David Fravor, observed a white, oval-shaped object that was eventually dubbed as the "Tic Tac."

This unique craft was able to demonstrate flight characteristics far beyond any known technologies. It was remarkable for its rapid acceleration, abrupt changes in direction, and the complete absence of any visible means of propulsion.

Crucially, these UAP sightings were fully corroborated by other naval aviators, by multiple sensor systems, including the specialized radars on the USS Princeton. The accounts were bolstered by the heads-up display recordings from the F/A-18s' cameras, including their advanced Forward-Looking Infrared (FLIR) cameras. The U.S. Department of Defense later confirmed the authenticity of the video footage. In 2020, the Pentagon formally acknowledged the release of three UAP videos as being factual, including the vital footage from the Nimitz incident.

The 2014 and 2015 USS Roosevelt East Coast Encounters

In 2014 and 2015, U.S. Navy aviators flying off the eastern coast of the United States reported repeated analogous encounters with UAPs. These strange objects appeared on radar, infrared systems, and were confirmed visually. They exhibited extraordinary capabilities through their amazing maneuvers. Several pilots reported near mid-air collisions, prompting them to create official safety reports. The "Gimbal" and "Go Fast" videos, which capture some of these incidents, were released publicly and have since been the subject of intense analyses from a variety of sources.

3. Declassified Government and Intelligence Reports

In recent years, the U.S. government has taken unprecedented steps toward transparency regarding UAPs. In June 2021, the Office of the Director of National Intelligence released a report analyzing 144 military UAP encounters identified between 2004 and 2021. The ODNI report concluded that most of the incidents remained unexplained, stating:

- In 143 of the 144 cases, analysts could not definitively explain the objects' nature or origin.

- Some objects appeared to demonstrate advanced technology, such as rapid acceleration, no discernible means of propulsion, and the ability to operate in multiple environments, including the air, the sea, and possibly even in space.

- There was no evidence released suggesting these objects were part of any U.S. secret program or foreign adversary technology, as was officially assessed by the American intelligence community.

Project Condign (UK)

The British Ministry of Defence conducted its own investigation, known as Project Condign. They analyzed thousands of UFO reports and released their findings in 2006. While often cautious in its conclusions, the report acknowledged that "Unidentified Aerial Phenomena" exist. They stated that some of these cases simply could not be explained by any conventional means.

4. Credible Civilian Sightings

While military cases dominate recent discussions, several civilian encounters have withstood decades of scrutiny and remain unexplained.

❖ The Kenneth Arnold Sighting (1947)

We will now take a deeper study regarding this crucial event.

On the clear afternoon of June 24, 1947, he was piloting a routine flight in Washington State. The routine was disturbed when Arnold reported seeing something extraordinary. His strange sighting would ignite a nationwide frenzy. Many today believe that he launched the modern era of UFO interest.

The Man Behind the Sighting: Kenneth Arnold

Kenneth Arnold was not a man given to flights of fancy. Born in 1915 in Sebeka, Minnesota, Arnold was an accomplished aviator and a successful businessman. He owned the Great Western Fire Control Supply, a company dealing in fire suppression equipment, and was well-respected in his community. As a skilled private pilot with over 4,000 hours of flying experience, Arnold's reputation for reliability and honesty made his account all the more compelling to the public and to the press.

The Events

The fateful day was June 24, 1947. Arnold was flying his CallAir A-2 plane from Chehalis to Yakima, Washington, on a business trip. At approximately 3:00 PM, as he approached Mount Rainier, Arnold noticed a strange, bright flash in the sky. At first, he thought it might be sunlight reflecting off another aircraft. However, as he looked more closely, he realized that what he was seeing was far from ordinary.

Arnold was able to observe a chain of nine unusual objects flying in formation and at remarkable speeds. He later estimated their speed at over 1,200 miles per hour. This was faster than any known aircraft of that time. The curious objects appeared to be crescent-shaped or disc-like, with a shimmering, metallic appearance. Arnold described their movement as erratic, "like a saucer if you skip it across the water."

"Flying Saucers" Enter the Lexicon

When Arnold reported his experience to authorities and the press, his description of the objects' motion, skipping "like saucers," was quickly paraphrased in headlines as "flying saucers." Though Arnold did not claim the objects were actually saucer-shaped (he described them more as boomerang or crescent-shaped), the phrase captured the public's imagination. His description soon became synonymous with unidentified flying objects.

The Reaction: Press, Public, and Military

Arnold's sighting was reported in newspapers all across the United States almost immediately. Given Arnold's credibility, his account was taken seriously. This prompted both curiosity and some degree of alarm. Within days, reports of similar sightings began flooding in from around the country. The U.S. military, already on alert in the postwar era, launched its own investigations into what had been seen that day over Mount Rainier.

Arnold's Testimony

Arnold was interviewed numerous times by journalists, military officers, and fellow pilots. He consistently maintained his account, providing detailed sketches and descriptions. He insisted he had seen something real and tangible. He asserted they were not any type of hallucination, and were not any "tricks of the light." Arnold's straightforward demeanor and his lack of sensationalism lent his account a gravity that immediately fueled further investigation.

The Birth of a Phenomenon

The Kenneth Arnold sighting did more than simply intrigue the public. Instead, it marked the beginning of the UFO phenomenon as we know it today. In the weeks following his report, hundreds of similar accounts surfaced, creating what became known as the "flying saucer craze" of 1947. The incident set the stage for ongoing public fascination and concern about unidentified aerial phenomena that would persist throughout the 1950s.

Government Investigations

In the wake of Arnold's sighting, the U.S. government launched formal investigations into UFO reports. The most notable of these early efforts was Project Sign, established by the Air Force in 1947. Project Sign sought to determine whether the objects reported posed any credible threat to national security. Nonetheless, Project Sign's findings were largely inconclusive. No answer would be brought before the American public.

Instead of solving the enigma, this questionable report spurred on even more interest. Trying to do a better job this time, Project Grudge was launched in 1949. It gave no further answers to the worried public. The Air Force decided to try for a third time, creating the infamous Project Blue Book in 1952. While the Air Force continued to investigate, the UFO phenomena grew ever stronger.

Scientific and Skeptic Perspectives

Arnold's sighting, while widely reported, was not without skeptics. Some suggested that what he observed could have been a mirage, a flock of birds, or even distant reflections of snow on the surrounding mountain peaks. Others hypothesized that Arnold may have seen experimental military aircraft, though there were never any records of such craft in the area at the time. Since then, there have still been no evidence of the true source of his sightings ever produced.

❖ The Phoenix Lights

On the night of March 13, 1997, thousands of people across Arizona, and notably over the city of Phoenix, witnessed a series of mysterious lights in the sky. The event, now famously known as the "Phoenix Lights," has since become one of the most discussed and enigmatic UFO sightings in modern American history. More than two decades later, the truth about what really happened remains a subject of intense debate, investigation, and enormous fascination. This case is a remarkable event to this day. No comprehensive overview of the Phoenix Lights event, examining eyewitness accounts, official explanations, alternative theories, and the enduring legacy of that extraordinary night, has successfully explained whatever it was that happened.

Chronology of the Event

The phenomenon began just after 7:00 PM and continued until approximately 10:30 PM, spanning several hours and covering hundreds of miles. The first reports originated in Henderson, Nevada, where a witness described seeing a V-shaped formation of lights heading southeast. As the evening progressed, multiple observers from Prescott and Dewey, Arizona, described similar sightings. Some of the people witnessing this event described a massive, silent craft with lights that seemed to float slowly through the night sky. By 8:30 PM, the spectacle reached Phoenix, where hundreds if not thousands of residents, from casual stargazers to entire families, stood transfixed as the lights moved overhead before them.

Descriptions varied, but most witnesses reported seeing a wedge or boomerang-shaped object. It was intermittently described as a solid dark mass that blocked out the stars as it passed. It was adorned with five to seven lights along its leading edge, as most observers reported. Others saw a series of orbs or lights moving in a straight or V-shaped pattern. While the event culminated in Phoenix, reports of these unusual light formations continued south toward Tucson.

Eyewitness Testimonies

The strength of the Phoenix Lights legend lies in the sheer number and diversity of witnesses. People from all walks of life, including police officers, pilots, scientists, and everyday citizens, all reported seeing something extraordinary. Many described a chilling silence accompanying the spectacle, as the mysterious object(s) moved with unexpected slowness and grace.

Some accounts mentioned that it was a massive craft, so enormous that it literally blotted out large patches of the stars above. This gave those observers a sense of an enormous scale for a vehicle that totally dwarfed any known aircraft.

Parents described children who were both frightened and mesmerized. Some recalled the communal amazement as groups of neighbors gathered in the streets, pointing skyward. For many, the event was transformative, turning skeptics into believers and prompting a lifelong interest in unexplained aerial phenomena. Notably, then-Governor Fife Symington initially downplayed the incident, but later admitted he too had witnessed the lights and called the experience "otherworldly."

Official Explanation

In the wake of widespread public attention, military and government authorities quickly offered their explanations for the Phoenix Lights. The United States Air Force stated that the second wave of lights observed later in the evening was due to a training exercise at the Barry Goldwater Range, southwest of Phoenix. According to their account, A-10 Warthog aircraft dropped a series of LUU 2B/B illumination flares, which slowly descended under parachutes. This then created a string of bright lights that would eventually disappear as they fell behind the Sierra Estrella Mountain range.

This official explanation was supported by video analysis from the evening. Footage of these later lights does indeed show them appearing to hover and gradually winking out, consistent with flares sinking behind distant mountains. However, this only accounted for part of the phenomenon. It specifically explained only the lights seen after 10:00 PM.

As for the earlier V-shaped formation observed traversing the state, authorities dismissed it as a formation of military jets flying together in formation, possibly using navigational or unique formation lights. However, this explanation failed to convince many, given the scale, silence, and slow speed described by so many of the witnesses.

Alternative Theories

Despite official statements, alternative theories have flourished, fueled by apparent discrepancies in timing, eyewitness accounts, and the magnitude of the event.

- Experimental Military Aircraft: Some believe the massive, silent craft was a secret military project, such as a stealth blimp or advanced aircraft being evaluated by the U.S. government. The American Southwest, with its proximity to a number of military bases and a long history of experimental aviation, provided a plausible context for such a theory. It was from Luke AFB that I learned to fly F-16 fighter jets. However, no concrete evidence or declassified documents have ever confirmed this to be the cause of the phenomenon.

- Extraterrestrial Visitation: For UFO enthusiasts and a great many other witnesses, the Phoenix Lights remain among the strongest evidence of alien visitation. The scale and behavior of the object(s), along with the lack of sound and the inability of conventional explanations to fully account for all sightings, have convinced many citizens that what was seen that evening was not of this Earth.

- Mass Hallucination or Misidentification: Some skeptics argue the event was just a case of mass misperception, with people misidentifying mundane aircraft, flares, satellites, or even migratory birds illuminated by city lights. However, the consistency and quantity of reports strongly counter this assertion for many analysts.

- Hoaxes or Pranks: While some have speculated that a group of pranksters could have flown aircraft in formation or released balloons fitted with lights, the wide scale and coordination required to make this explanation viable seems far less credible.

This event also played a role in shaping modern discussions about "UAP" events. In recent years, the U.S. government has taken a number of steps to acknowledge and investigate similar reports more seriously. This suggests there is a shifting attitude toward unexplained sightings. We can hope they truthfully follow such a plan.

❖ The O'Hare International Airport Incident

On November 7, 2006, Chicago's O'Hare International Airport became the focus of worldwide attention. This was not for a new route or any record-breaking passenger numbers. Instead, it was because of an unexplained aerial phenomenon that would spark debate, speculation, and investigation for years to come.

The O'Hare International Airport UAP incident remains one of the most compelling and credible reports of an unidentified flying object in recent history. We need to document the events of that day. There were witness accounts, then subsequent federal and corporate investigations, and the broader implications for considering public discourse on UAPs in general.

The Incident: A Moment Over Gate C17

The extraordinary events took place on a typical overcast afternoon. At approximately 4:15 PM, United Airlines employees working at Concourse C, specifically near Gate C17, noticed a curious and alarming sight. Above the terminal, a metallic, disc-shaped object hovered silently in the low clouds. What began as a sighting by a single ramp employee quickly grew into a chorus of astonished voices, including pilots, airport workers, and aircraft mechanics. All of these were seasoned flight professionals and were used to seeing aircraft of all kinds. Yet on this day, they still gathered to observe this unusual phenomenon.

Multiple witnesses described the object as a gray, rotating disc, estimated to be about 6 to 24 feet in diameter. It hovered for several minutes, seemingly defying the laws of conventional physics. There were no observable means of propulsion, no flight controls, no antennae, no sounds to be heard, and no visible lights, even as the object remained steady in place, seemingly indifferent to the wind and weather.

Eyewitness Accounts

What sets the O'Hare incident apart from many other UAP reports is the quality and credibility of the witnesses. United Airlines employees, including ground crews, pilots, and management personnel, provided their consistent and detailed descriptions. They recounted the object as a perfectly smooth, metallic craft resembling a "saucer" or "disc," distinct from any planes or helicopters in service.

A pilot taxiing for departure caught a glimpse and, via his radio, asked the control tower if they had visual or radar confirmation. However, the object did not appear on the Federal Aviation Administration radar, nor did it show up on any FAA airport tracking equipment.

Witnesses described the object's sudden departure as equally remarkable. After several minutes of it remaining perfectly still, it suddenly shot upwards at a tremendous speed. Many people reported seeing it punch a clean, circular hole through the cloud cover, which lingered for several minutes. Of interest is the fact that no sonic booms were reported by any witnesses, considering the obvious speed used by the craft.

Initial Response and Official Reaction

Despite the number and reputation of the witnesses, United Airlines did not publicly acknowledge the incident at first. Employees were informally cautioned not to discuss it with the media. The FAA initially dismissed these reports, attributing the sightings merely to "weather phenomena" or misidentified natural occurrences.

However, due to persistent inquiries, particularly from the Chicago Tribune and its transportation reporter Jon Hilkevitch, documents and audio recordings obtained via the Freedom of Information Act eventually revealed that a number of FAA officials discussed the incident within their own internal communications. While the FAA officially maintained that the sighting posed no safety concern and was likely a weather effect, leaked documents from the agency suggested a much more serious level of internal attention.

Media Coverage and Public Interest

The O'Hare incident quickly garnered national and international coverage, including by mainstream outlets such as CNN, CBS, and the BBC all reporting on the story. The credibility of the witnesses, many of whom had years of aviation experience, truly added to the intrigue. The story reignited public debate about UAPs, the lack of governmental transparency, and the urgent need for serious scientific investigation of unexplained aerial phenomena.

The FAA's Investigation and Criticisms

The lack of physical evidence, with no photos or videos that were made public, and the fact that the object left no material trace behind, aside from the reported hole in the clouds, clearly hindered any thorough investigation. The FAA's refusal to investigate further, combined with its dismissal of the event as simply a weather phenomenon, drew intense criticism from both the public and some aviation professionals.

The Witnesses Speak Out

In the years following the incident, several witnesses have spoken publicly, either anonymously or under their real names, in interviews, documentaries, and podcasts. Their testimonies have remained consistent. They say that the craft was unlike anything they had ever seen, and that its behavior could not be explained by any conventional technology or natural phenomena. Due to this, the sense of intense frustration was common among the professional witnesses and UAP enthusiasts.

❖ **There are So Many Other Cases Where Physical Traces or Materials Have been reported:**

The phenomenon of UAPs has intrigued humanity for decades. While many sightings involve only visual encounters, some cases present compelling physical evidence, known as "landing impressions." These are marks or changes on the ground allegedly created by the landing or close approach of a UAP. Such impressions range from scorched vegetation and depressions in the earth to strange residue, altered soil chemistry, and even electromagnetic anomalies. Here, we explore notable examples, the types of traces reported, and the ongoing debate over their origins.

What are UAP Landing Impressions and Deviations?

UAP landing impressions are physical alterations to the environment where an alleged UFO has landed, hovered close to the ground, or otherwise interacted with the terrain. They can take many forms, including circular or oval depressions in soil, grass, or crops. There can be burned, singed, or dehydrated vegetation at the site. Branches of trees have been broken by the craft. There have been noteworthy changes in the soil composition and even increased radiation levels noted, deforming the very location of interest. In some instances, there were unusual residues or substances left behind

Magnetic or electromagnetic disruptions have also been found. Such effects of compasses and electronic equipment are often documented by UAP investigators, interested scientists, and sometimes by law enforcement personnel. These provide tangible evidence for researchers and are frequently the subject of both fascination and skepticism.

Famous Examples of UFO Landing Impressions

1. The 1964 Socorro, New Mexico Incident

On April 24, 1964, police officer Lonnie Zamora witnessed a strange egg-shaped craft land in a ravine near Socorro, New Mexico. After observing two small figures and a loud roaring noise, Zamora saw the object lift off and depart. When he approached the landing site, he found:

- Four shallow, V-shaped impressions in the ground, forming a rough rectangle
- Burned and scorched brush
- Small footprints and metallic residue

The Socorro incident was investigated by Project Blue Book and remains one of their best-documented cases with the findings of physical traces left behind.

2. The 1980 Trans-en-Provence Case, France

On January 8, 1981, a farmer named Renato Nicolai reported seeing a disc-shaped object land on his property in Trans-en-Provence, France. The craft remained for less than a minute before taking off. Investigators from GEPAN (a French government UAP study group) found:

- A circular trace in the soil, approximately 2.4 meters in diameter
- Pressed vegetation and altered chlorophyll levels
- Changes in soil composition, including the presence of elevated amounts of zinc and phosphate at the site

This case is notable because of the scientific analyses conducted and the official recognition by French authorities.

3. The 1966 Tully Saucer Nest, Australia

On January 19, 1966, banana farmer George Pedley claimed to have seen a gray, saucer-shaped object rise from a swamp in Tully, Queensland. He discovered a "nest" to be there.

- It demonstrated a flattened, circular area in the reed beds, about 9 meters in diameter
- Vegetation was bent and interwoven in a clockwise direction
- No evidence of burning, but the formation remained visible for weeks

- Multiple similar nests were later reported to have been found in the area, sparking debate over their origin.

4. The 1971 Delphos, Kansas "Glowing Ring"

Teenager Ronald Johnson and his family reported seeing a mushroom-shaped object land on their farm on November 2, 1971. At the site, they found:

- A glowing, ring-shaped impression on the ground
- The soil was water-repellent and had a peculiar, crusty texture that would no longer grow plants
- Laboratory tests that showed chemical changes and fungal spore alterations where the "landing" had been

The Delphos case is considered one of the most unusual, with the "glowing ring" observed by multiple witnesses over the coming days.

5. The 1979 Valensole Encounter, in France

French farmer Maurice Masse claimed to have seen a small spherical craft and two humanoid figures set down in his lavender field on July 1, 1965. After the encounter:

- He found a circular impression with small holes, as if left by landing gear
- The lavender plants in the area reportedly died and wouldn't regrow for years
- Masse suffered from unexplained fatigue and recurring sleep disturbances after he examined the area of the event

- Investigators documented the traces and interviewed Masse, who maintained his story for decades.

Common Features of Landing Impressions

Though each case is unique, many share certain characteristics:

- Circular or geometric patterns, often with symmetry suggesting mechanical origin
- Soil compaction or displacement, sometimes with increased radiation
- Vegetation that is scorched, dehydrated, or otherwise damaged in a manner inconsistent with known machinery. It is common that these locations have persisting difficulties in growing plants afterward.
- Traces that often appear overnight or in remote areas, with no clear evidence of conventional vehicles.
- Some cases also document trace materials, such as powders, gels, or metallic residues, even though later analyses frequently yield inconclusive results.

It is noteworthy that these same types of deformations are also linked to many crop-circle cases.

Scientific Investigations

Scientific scrutiny of UFO landing impressions varies. Some traces have been subject to rigorous laboratory analysis, revealing anomalies such as:

- Changes in soil pH and mineral content.

- Degeneration or alteration of plant cellular structures.
- Unusual isotope ratios or trace elements.

However, skeptics point out that many impressions can be attributed to natural or mundane causes such as:

- Animal activity (e.g., nests or sleeping spots).
- Weather phenomena (e.g., lightning strikes or whirlwinds).
- Human activity (e.g., vehicles, hoaxes, or agricultural equipment).

Despite these possible explanations, many cases remain unexplained and are considered to exhibit "high levels of strangeness" events within UFO literature.

Eyewitness Testimony and Its Role In UAP Investigations.

Here are some common elements humans experience in these settings:

- Landing impressions gain credibility when accompanied by consistent eyewitness accounts.
- Bright lights or craft seen descending or resting on the ground.
- Loud noises, electromagnetic interference, or strong winds. To the opposite end, some craft that were close enough that sounds should have been heard were said to be completely silent.
- Unusual smells or sensations noted by witnesses.
- Physical effects on people or animals, such as temporary paralysis, agitation, skin effects, and difficulties with their sleep.

While eyewitnesses are susceptible to error and suggestion, the frequent correlations of physical and testimonial evidence strengthens many particular cases. With the events happening to so many people in so many different locations, it is hard to believe that there is no validity to these reports.

6. Historical Art, Photographic, and Video Evidence of UFOs

The tradition of reporting mysterious aerial phenomena dates back centuries, with ancient art and chronicles describing wondrous lights and objects in the sky. Skeptics will claim that medieval artists who painted UFOs in their works were not painting literal interpretations of reality but rather symbolic narratives. Yet, some works from this timeframe clearly feature mysterious flying objects that modern viewers would definitely associate with UFOs.

- The Baptism of Christ (Aert de Gelder, 1710): Though from the early modern period, this work is frequently cited in UFO literature. It shows a radiant, disk-shaped object in the sky beaming light down upon Christ. Many other medieval and Renaissance paintings similarly depicted circular clouds or mandorlas as symbols of divine presence. But for everyday UFO enthusiast of today, their modern resemblance to UFO imagery is particularly striking.

- The Madonna with Saint Giovannino (Domenico Ghirlandaio, circa 15th century): This painting, remaining in Florence, shows the Virgin Mary with a highly mysterious object in the sky behind her. To the right, a man and his dog gaze upward at the strange, disk-shaped form. While non-believers argue the object likely represents angelic or divine presence, it has become iconic among UFO enthusiasts.

- The Crucifixion (Kosovo, 14th century): In the Visoki Dečani Monastery, a fresco of the Crucifixion shows two enigmatic, pod-like objects on either side of Christ on the cross. Remarkably, within each of these flying objects sits a human figure inside. Some skeptical viewers interpret these as merely representations of the sun and moon or as angelic hosts. For UFO enthusiasts, their stylized forms invite a clear comparison to spacecraft, as we would identify them in our day.

- "The Annunciation with Saint Emidius" (Carlo Crivelli, 1486): In this painting, a beam of intense light, similar to a laser, emanates from a circular opening in the clouds. It strikes the Virgin Mary directly. The opening in this cloud is sometimes interpreted as a celestial object or portal, though it most likely symbolizes the Holy Spirit. For people of our day, many would conclude that it is easily recognized as a UFO episode.

The first widely publicized photograph of an alleged UFO was taken back in 1870, showing a cigar-shaped object. It was noted to be hovering above Mount Washington in New Hampshire. However, it was the 20th century, with its proliferation of cameras and, later, video recorders, that brought a new dimension to the discussion. Suddenly, people could not only recount their sightings, but also attempt to preserve them visually using these recording methods.

- The McMinnville Photographs of 1950

One of the most famous instances of alleged UFO photography occurred in McMinnville, Oregon. Paul and Evelyn Trent claimed to have seen a flying object near their farm and snapped two iconic photos. These images, showing a disc-shaped object in the sky, have been scrutinized for decades. Analyses by photographic experts and skeptics have alternately declared them to be authentic, and yet also dismissed them as hoaxes. Despite their age, these photos are often cited by UFO enthusiasts as among the most compelling.

- The Lubbock Lights of 1951

In Lubbock, Texas, a series of photographs taken by college students captured bright lights in a V-formation crossing the night sky. Witnessed by multiple people, the phenomenon remains unexplained, though some suggest a flock of birds reflecting city lights was responsible. The case demonstrates how photographic evidence can both support and complicate eyewitness testimony.

- The Hessdalen Lights of Norway

Since the 1980s, the remote Hessdalen valley in Norway has been the site of recurring unexplained lights. These have been documented by numbers of still cameras, interesting videos, and even scientific monitoring stations. Some images and footage show bright, fast-moving orbs, occasionally splitting or changing colors. The remarkable visual effects have attracted scientific attention from around the world. These lights remain a rare example of a persistent, documented phenomenon. Their reports continue to occur even today.

- **The Phoenix Lights**

As we have noted, on March 13, 1997, thousands of people in Arizona reported seeing a vast formation of lights over the city of Phoenix. Numerous videos and photographs were taken, showing luminous orbs in a triangular pattern. Explanations have ranged from military flares to extraterrestrial craft, and the event is often cited as a modern classic of UFO evidence.

The Challenge of Verification

Film photography, and now digital video technology, have empowered many to document what they have seen down through the years. However, these same technologies are susceptible to various limitations. Effects such as low resolution, poor lighting angles, camera shake, and, increasingly, intentional digital manipulation are to be considered with these images. Many alleged UAP images have been debunked as merely camera artifacts, misidentified natural or man-made objects, or outright fabrications created by intentional hoaxers.

The Difficulty of Context

Unlike eyewitness accounts, photographic and video evidence can be analyzed repeatedly. Yet, without reliable provenance and contextual information, such as time, location, and witness testimony, images can still be difficult to authenticate. In many cases, even experts disagree on whether a given piece of evidence is credible.

The Role of Modern Technology

Today, nearly everyone carries a smart phone camera in their pocket, and social media platforms can spread images instantly around the world. This has led to a dramatic increase in reported sightings and purported evidence. However, the sheer volume of material makes it challenging to separate genuine anomalies from deliberate fakes or innocent mistakes.

Image Analysis and AI

Advances in image analysis, including artificial intelligence and machine learning, have improved the ability to detect manipulated images and videos. Researchers can now examine metadata, search for telltale signs of editing, and compare new footage with known phenomena. Despite these tools, the ambiguity inherent in many UFO images persists.

In recent years, the U.S. government has revived interest through the Unidentified Aerial Phenomena Task Force and the All-Domain Anomaly Resolution Office. The actions of these agencies, nevertheless, have not proven reliable to many UAP believers. Their released videos have instead prompted further calls for additional transparency and greater scientific inquiries, though answers to these criticisms remain elusive.

Cultural Impact of Photographic and Video Evidence

Sensational UAP videos and photographs have had a profound influence on movies, television, books, and art. Scenes of flying saucers and mysterious lights have become visual shorthand for the unknown, inspiring generations of creators and believers. Just consider the impact of the Star Wars® and Star Trek® movies on our shared culture. May the Force be with all Trakkies!

7. Testimonies from High-Level Officials

Over the years, numerous government and military officials have come forward to state that some UAP incidents remain unexplained. This would include:

- Former U.S. Senator Harry Reid, who championed the Advanced Aerospace Threat Identification Program (AATIP), stated: "Much of what we observed remains unexplained, and there is compelling evidence that we may not be alone."

- Astronaut Edgar Mitchell was the sixth person to walk on the moon during his Apollo 14 mission. While Mitchell never openly claimed to have seen a UFO while in space, he stated in a number of interviews that he was briefed by high-ranking officials who confirmed that extraterrestrial craft have visited Earth. His stature as a scientist and astronaut adds gravitas to his claims, even if they are based solely on second-hand accounts.

- Gordon Cooper, an Oklahoma native, was one of the original Mercury astronauts and also flew with NASA's later Gemini Program. Cooper reported that he had seeing UFOs while piloting an F-86 over West Germany back in 1951. He described them as silver-colored, saucer-shaped objects that outmanoeuvred any known technology at the time. Cooper also claimed that, in 1957, a UFO had landed at Edwards Air Force Base and was actually filmed by a camera crew that he supervised.

- Astronauts Frank Borman & James Lovell, Jr.

During the space mission of Gemini 7 in 1965, astronauts Frank Borman and James Lovell Jr. reportedly identified a strange sighting that ufologists still believe was a real UFO. In the transmission, it was referred to as a "bogey." This is military speak for something unknown. Borman claimed it to be a separate spacecraft. Their communications were cut off for a short time. Once the mission ended, Borman claimed that it was not a real UFO but a simple mistake. Who knows what these two astronauts actually witnessed?

- General Wilfried de Brouwer was a senior Belgian Air Force official. General de Brouwer was heavily involved in investigating the 1989–1990 Belgian UFO wave. Multiple sightings were corroborated by radar, police, and Belgian military personnel. De Brouwer held several press conferences and eagerly provided evidence that the mystery UFOs had displayed flight characteristics beyond any technology that he knew about.

- Dan Aykroyd is well-known for his role in "Saturday Night Live" and "Ghostbusters." He has been a vocal believer in UFOs. He claims to have witnessed UFOs and has produced documentaries on the subject, drawing attention to lesser-known cases.

- In a case that will be reviewed in more detail later, future president Jimmy Carter witnessed a UFO and even reported it to the International UFO Bureau in Oklahoma City. His report is available both at the Bureau and at his presidential library. I must admit that I learned about this case when I joined the Bureau myself. I will declare that I am a Member of the Board for this agency.

- High-ranking military officers, pilots, and even intelligence officials from multiple countries have continued to affirm the persistence of unexplained aerial phenomena.

8. The Search for Answers Continues

The current best evidence for real UAPs is not in any single case or piece of footage, but in the cumulative weight of the unexplained phenomena witnessed and documented over so many decades.

The U.S. government, especially Congress and the scientific communities, are seemingly now more open to engage this subject. They appear to be seeking to distinguish prosaic explanations from genuinely extraordinary events. As data collection improves and public transparency increases, the world may be on the cusp of breakthroughs in the understanding of these enduring mysteries.

One More of My Personal Stories

My NASA/DDMS Deployment to the Ben Guerir Air Base in Morocco

The Ben Guerir Air Base is located out in the baking sun of the Marrakech-Safi region of Morocco. While most Americans have never heard of it, the base is a military site that has played a key role at various points in U.S. military and aeronautical history. With its modern infrastructure and strategic location, it remains notable for its importance in international relations and defense collaborations.

Origins and Development

The Ben Guerir Air Base was originally designed to meet military needs during times of the Cold War global tensions. Situated about 60 kilometers north of Marrakech, its geographical position made it ideal for a variety of military operations. The site was developed with long and robust runways capable of accommodating large aircraft, including strategic bombers and transport planes.

During the peak of the Cold War, the Ben Guerir Air Base was utilized by the U.S. B-52 bombers, capable of flying their nuclear payloads into the Soviet Union. Its location in North Africa was essential for surveillance operations and providing logistical support for potential international bombing missions. It was part of the large network of bases established by the United States' Strategic Air Command (SAC) to counter the Soviet influence that had been growing within the Middle Eastern region.

The base, however, is also renowned for its role in the American space program. It was used as a Transoceanic Abort Landing (TAL) emergency landing site for the space shuttle program. This was precisely why I travelled over to Morocco in 1992 while I was stationed at Patrick AFB and Cape Canaveral AFS.

The Department of Defense Manned Spaceflight Support Office was the oversight organization for the military support of NASA's manned spaceflight operations all over the world. You see, there were many requirements for DDNS to ensure that the space program was prepared for all shuttle contingencies. It took much effort, all around the globe, to confirm that the DoD could effectively support whatever actions were required to safely protect an orbiter and crew that landed outside of American territory.

At Ben Guerir, as at the other three TAL sites, crews would have to be present and fully prepared just in case anything went wrong with a space shuttle launch. Security teams were mandatory for the safety of the astronaut crews and other American support personnel. There had to be highly knowledgeable ground crews available because once a space shuttle orbiter would land, there were an amazing number of tasks that would be required. In case of a crash, hazmat teams would be extremely valuable. Rescue crews would need to brave terrible hazards to remove the crew to safety. Flight surgeons and medical teams would have to be present to provide whatever emergency medical care would be required to treat any of the injured astronauts.

Because crews supporting each TAL site might be gathered from military facilities all across Europe, the flight surgeons would need training and educational materials to teach them how to support a shuttle operation ending on the wrong side of the Atlantic Ocean.

I happened to be the author of the TAL Support Supplement to the Space Transportation System (STS, or space shuttle) DoD Medical Personnel Guide. This handbook was provided by DDMS to all military medical providers with the possibility of being involved with such a mission. Because of this, I was at Ben Guerir to inspect the operations there.

Military personnel, excluding MI-6 operators like James Bond, do not take any fancy clothing when they deploy overseas. My guys and I mainly had our flight suits, blue jeans and associated common clothing. One afternoon we learned that the Moroccan Air Force liaison had arranged for us to eat at a very upscale dining restaurant in Marrakech. We could not pronounce his actual name, so we all called him Bob Freeway. Cute?

We were staying at the Hotel Atlas in Marrakech, which was very nice, and was not far from where we would be eating. Bob assured us that we could dress as we could. When we arrived, we saw that the other diners were in suits, formal dresses, and a few even wore tuxedoes. We were the only visitors in blue jeans. They thankfully told us they would also give us discounts on the price of our meals. That was cool with me!

When we got there, Bob had told them that an astronaut would be with us. I won't name the astronaut, but he was a really fine down-to-earth kind of guy. When our group arrived, there was one man who had insisted in wearing his security badges around his neck, just to show off his own self-importance. He had even brought several security badges from KSC with him to Africa, I guess to look more impressive. He was the ground operations manager (GOM) and usually spoke louder than necessary anyway.

Since the astronaut was dressed just like the rest of us and the GOM had his "necklace of great importance" on, the host went to him, expecting him to be the astronaut. Naturally, the GOM really ate up the attention he was getting while the true astronaut went over to the table they had for the rest of us and sat down. The GOM, a.k.a. fake astronaut, was taken to a special table and was being served by two waiters.

One of our PJs (pararescue jumpers) asked the astronaut if he was going to tell the waiters that he was the real astronaut. He replied that he would rather sit with us. Throughout the entire meal, the GOM was eating up the attention while the astronaut hung out with us. The waiters, other restaurant employees, and even some of our fellow diners had their pictures taken with the GOM.

Finally, as we were preparing to leave, the owner of the restaurant and the chefs from the kitchen came out to have their pictures taken with "the astronaut." Things took a wrong turn for the GOM when the owner asked to put his picture on the wall and asked when he would next fly in space. At that point, the GOM pointed over to the astronaut, admitting the truth.

The owner and head chef rushed to the astronaut and began to profusely apologize for treating him as a regular-type diner. The actual astronaut was very gracious and told them there was nothing to apologize for. After all, he had a terrific meal and was treated well, as had we all.

The owner and head chef then took the authentic astronaut over to the large fireplace they had in place on the far wall and had a bunch of photos taken with him. As I understood, they did not charge the astronaut for his meal and the rest of us only had to pay half-price for our meals as well.

On the drive back to the hotel, none of us spoke to the GOM. We figured he had already received all of the attention he deserved for that evening!

We Got it From Them!

Chapter 7

The Search for Extraterrestrial Intelligence (SETI) and UFOs

The search for life beyond Earth is a robust area of scientific inquiry. Projects like SETI survey the cosmos for radio signals or optical beacons from alien civilizations. While UFO sightings are sometimes linked to theories of extraterrestrial visitation, the SETI community typically takes care to distinguish between scientific searches and anecdotal UFO reports.

Many physicists have debated the feasibility of interstellar travel. The vast distances involved make routine visits from other star systems unlikely, with our currently understood technology. However, some UFO believers propose that aliens might have advanced propulsion methods, such as wormholes or warp drives, can manipulate physics in fashions we do not recognize, while others suggest the craft that we witness might be merely probes, rather than crewed ships.

But what has SETI actually uncovered?

SETI's mission centers on detecting evidence of extraterrestrial intelligence through the observation and analysis of electromagnetic signals, primarily in the radio and optical spectra. The underlying premise is that other intelligent civilizations, if they exist, may use technology similar to ours and might inadvertently or deliberately emit signals into space that could be detected from Earth.

Radio Signal Detection

The most iconic approach of SETI involves scanning the skies for narrow-bandwidth radio signals. These would be emissions that are unlikely to be produced by natural astrophysical processes. Large radio telescopes, such as Puerto Rico's Arecibo Observatory (before ending its useful life), the Robert C. Byrd Green Bank Telescope in West Virginia, the Very Large Array (VLA) radio telescope facility located 50 miles west of Socorro, New Mexico, and many others, have served as the primary instruments for these surveys.

Optical SETI

In addition to radio SETI, researchers have begun searching for brief, powerful bursts of light (optical SETI) which could be indicative of laser communication used by advanced civilizations. These would be harder to discern, but the scientists want to leave no stone unturned.

Technosignature Searches

SETI also investigates so-called "technosignatures," which include not just artificial radio or optical signals, but also any potential evidence of advanced engineering. This might be in the form of Dyson spheres, any unusual chemical signatures in distant planetary atmospheres, or any anomalous energy outputs from stars.

Summary of Evidence Uncovered by SETI

To date, SETI has not uncovered any confirmed evidence of extraterrestrial intelligence. Despite decades of observation and analyses, no signal or phenomenon has passed the rigorous standards required for a declaration of contact or discovery.

Notable Signals and Events

While no confirmed evidence exists, SETI's history is marked by a handful of intriguing events:

- The "Wow!" Signal (1977): On August 15, 1977, Ohio State University's Big Ear radio telescope detected a strong, narrow-band radio signal that bore the expected hallmarks of an artificial origin. Astronomer Jerry R. Ehman famously annotated the computer printout with "Wow!" The signal lasted for 72 seconds but was never detected again, despite subsequent follow-up. Its origin remains unexplained, and while it ignited popular fascination, it cannot be considered evidence without replication or further data.

- Other Anomalous Signals: Throughout its history, SETI has detected various unusual signals. This would include the "Phoenix Project" candidate signals and unexplained blips during their META and BETA projects. However, these have always been attributed, upon further investigation, to terrestrial interference, satellites, or natural astrophysical sources.

- Repeating Fast Radio Bursts (FRBs): More recently, astronomers have detected mysterious bursts of radio energy from deep space, known as fast radio bursts. While most scientists believe these are natural in origin, the possibility of technological sources has not been entirely ruled out. SETI continues to monitor such phenomena, but none have provided the desired definitive evidence for extraterrestrial intelligence.

Challenges in the Search for Evidence

The absence of confirmed signals (excluding those found in the movie "Contact") is not entirely surprising, given the immense challenges facing SETI research. The universe is incomprehensibly large, with up to 400 billion stars just within the Milky Way galaxy alone. There are many more in the billions of additional galaxies beyond our own. SETI's telescopes can only listen to small portions of the sky and limited frequency ranges at any one time.

Signal Attenuation and Interference

Even if an extraterrestrial civilization is transmitting, their signals might be too weak to detect after traveling vast interstellar distances. Earth-based interference, from satellites and terrestrial sources, often complicates our own signal identification. Some scientists worry that interstellar space could contain enough disturbing radiation to totally disrupt any signals originating from stars so very distant from us.

The Fermi Paradox

The lack of evidence, despite the enormous number of potentially habitable planets, is encapsulated in the so-called Fermi Paradox: If intelligent life is common, why haven't we found any sign of it? Explanations range from the rarity of life and intelligence to the possibility that civilizations self-destruct or simply choose not to broadcast their presence. After all, no society can really know who might be out there listening!

Defining Evidence

SETI employs rigorous statistical and technical standards before declaring any discovery as evidence. It must be reproducible, should rule out all natural and human-made sources, and is required to be independently confirmed by multiple observatories.

Technological Evolution and Expanding Possibilities

As technology advances, so do SETI's methods:

- Expanded Wavelength Coverage: New telescopes and instruments allow SETI to monitor broader frequency bands, both in radio and optical spectra.

- Automated Data Processing: Machine learning and artificial intelligence are even now being employed to sift through massive datasets searching for any unusual or promising signals.

- Citizen Science: Projects like SETI@home have enabled millions of people to help analyze radio telescope data using their own personal computers and their volunteered time.

- International Collaborations: SETI efforts are global, with increasing cooperation between observatories and research groups worldwide.

Current Status and Future Prospects

SETI remains a vibrant and evolving field. Recent decades have seen the discovery of thousands of exoplanets, including some being found within the "habitable zone" of many star systems. These are planets where liquid water could exist, which would be a very encouraging sign for those hoping that life is common within the universe.

New Initiatives

The Breakthrough Listen project, launched in 2015, represents the most comprehensive SETI program to date. It uses state-of-the-art telescopes to survey nearby stars and galaxies across a broad range of target frequencies. While no evidence has yet been found, the scale and sensitivity of these searches continue to grow. Still, SETI scientists emphasize patience and the need for long-term, systematic searches.

The absence of evidence is not evidence of absence, which is certainly true for both SETI and the UAP fields. The universe may yet yield some amazing surprises as our technologies and observational reach is expanded.

A Story Once Again

While I was in medical school, I received payments for tuition, fees, books, and equipment, along with a monthly living allowance because I was a recipient of an HPSP (Health Professions Scholarship Program) military scholarship. Because of this, I received the rank of a second lieutenant and went on active duty for training each summer. During my final year in school, I was able to arrange a rotation to the emergency department at Tripler Army Medical Center.

This was quite a sacrifice for me because Tripler is located in Honolulu, Hawaii. I figured that if the Department of Defense truly needed me in Hawaii, I was willing to perform those difficult tasks I had been given. This is a joke, of course, since I had never been out to Hawaii and here I was getting paid to be there.

Tripler is a renowned medical facility that serves as a cornerstone for health services in the Pacific region. Known for its distinctive pink-colored architecture, this medical center stands out not only for its visual appeal but also for its extensive contributions to military medicine and community health. Its nickname was the pink lady on the hill.

I had all kinds of experiences there. One day a man came through the door with his hand clutching his chest and complaining that he couldn't breathe. Before anyone could do anything, he collapsed unconscious onto the floor. A resident-doctor saw this and called for a code, so this Code Blue was announced over the speaker system.

A group of doctors and nurses urgently rushed to the scene. We brought a cardiac monitor, which was set up immediately. There were nurses on each arm placing large-bore IVs so we could provide intravenous meds and fluids. While all of this was going on, the resident had started chest compressions for CPR.

About that time, this little old lady got out of her chair and went to this resident. She tapped him on the shoulder and announced, "I was here first!"

One evening while I was on duty, we got a call for an urgent ambulance mission. There had been a fight in a bar in Waikiki and one person was unconscious. We dispatched the ambulance and crew while we were taking care of the people already in the ER. A few minutes passed before the crew radioed in that they were on the scene. For probably at least another ten minutes, we heard nothing from the crew. This was unusual because they were good at giving us a heads-up about whatever situation had confronted them.

A few more minutes passed before the radio crackled to life. The driver had great tension in his voice as he told us, "Tripler, we are inbound to your location. Be advised that we have a female Marine patient, approximately six foot one and around 200 pounds. She got into a fight with three soldiers, one of whom made some comment she didn't like. She decked the guy. His two friends tried to intervene, and she was beating them up. One of the men took a chair and smashed it across her head. She became unconscious and remained so upon our arrival."

The radio voice went silent, but we could hear the sounds of a scuffle going on. The head ER doctor asked, "What's going on in there? It sounds like you're having a fight."

After another minute or so passed, the man radioed back, "Tripler, be advised that she has woken up and is tearing up the two medics in the back! Our ETA is three minutes, and we recommend that you have at least four security guards standing by! I repeat, we need security guards, and that is plural, standing by for when we pull in!"

As the ambulanced pulled up to the door reserved for their use, we had only two security guards, but we also had me, three male orderlies, two male residents, and this was being observed by the head ER doctor. We still were not ready for what awaited us!

When the back doors of the ambulance were pulled open, that woman burst out of them, followed by the two medics. She was angrily punching, kicking, biting, and spitting at us all. We finally had two guys on each arm, but we still could not control her legs. There was a wheelchair nearby so we managed to force her into the chair while six or eight of us continued the battle.

One of the orderlies begged the ER doc to give her a shot to knock her out, but he responded that she might already have a concussion or worse, so sedation was out of the question.

The only way we began to control her was to wrap her shoulders and arms with roll gauze. We had to keep placing more rolls around her arms and trunk before we felt as though we had her at least somewhat controlled.

Then it got worse!

With her legs still free, she noticed that the handle of the wheelchair was located at a very vulnerable spot for the security guard standing right behind her. She planted her feet on the floor and forcefully shoved that wheelchair backwards with all her strength. Sure enough, the handle struck her intended victim exactly as she had planned.

Enter foolish security guard number two. He stepped directly in front of this woman and insistently declared, "We will not put up with you acting like this!"

In a picture I shall never forget, a big smile came to her face as she kicked this guard in the same location she had injured the other guard!

The ER doc demanded, "Roll up her legs as well. We might as well make her look like a mummy before she kills someone."

That was what we did. By the time we were finished, we had wrappings from her ankles to the tops of her shoulders. She was, however, still spitting and cursing at us!

The ER doc turned to me, of all people, and said, "Young Doctor Rogers, you are now on her case. Show me how you can handle the situation."

Truthfully, this lady Marine was not a challenge I was yet ready for.

I let out a sigh, then pointed over to a treatment bed. I told the three orderlies I had available to me, "Let's get her up on that bed and I'll start to examine her."

With pure anger in her drunken eyes, she asked, "You think you can force me up into that bed?"

I determinately shook my head up and down. "We certainly will."

She spat at me and earnestly said, "Better men than you have tried to get me to bed!"

I looked to my orderlies and countered, "I don't believe that. Gentlemen, let's do this. Let's get her up there."

One thing I had not recognized was that with her totally rolled up as a mummy, there were no handles or anything else to hold on to. As the four of us tried to pull her out of the wheelchair and onto the bed, she was wriggling like a fish, which caused us to drop her onto the hard floor. I saw that she bumped her head pretty hard, so I didn't want to try that approach again.

I told my warriors, "Let's roll her onto a sheet and then lift the sheet onto the gurney."

After great effort, we finally got this woman onto the exam bed and lifted the bedrails into place. We were all fairly pleased with what we had accomplished. However, she was not through yet!

I instructed two medics to cut away the bindings on her right arm and then to tie the arm to the right guardrail. I thought to myself, "One extremity was tied down, but three more were needed."

Once this was done, I took one orderly on my side and two on the other side and we cut through the bindings on her legs. She could have been a soccer player with the kicks she was landing on us, but we finally managed to get each of her legs tied to the guardrails on either side.

Since all we lacked was the left arm, I turned the two orderlies on the other side back to other duties. I held firm to her wrist as the remaining orderly kneeled down to cut away the restraints on this arm. He was a strong African-American young man and he was cautious with his actions. It still took all of my force to hold her left wrist in place while this was happening.

Just then, from the corner of my right eye, I saw a sudden blur!

She had gotten her right hand loose, and it was trying to grab the Afro of my orderly. I immediately let loose of her left wrist while I latched onto her right arm and yelled, "Help!"

Perhaps four or five guys jumped in to assist me as the black orderly felt to make sure he had all his hair remaining. I can assure you; we needed every one of those orderlies to regain control of our highly intoxicated but extremely enraged patient.

When we were able to get a CT scan of the head and it appeared normal, we quickly transferred her to an inpatient room upstairs. We gave a very firm warning to that medical team what they were in for. After all, we had scrapes and cuts and bruises all over us from having to deal with her.

Not to mention the sad condition of the two security guards!

Years later, the Pentagon made the decision to allow women into combat zones. When I thought back to that lady Marine, I believed that was a good decision.

Chapter 8

The Societal Impact of UFOs

Beyond the expectations of pure science, UFOs occupy a significant place in our popular culture. They have inspired generations of films, literature, conspiracy theories, and even new religious movements. Encounters with the unknown can seriously challenge our cosmic perspectives, provoke fear, or wonder, and even influence our human technological development.

❖ The Influence of Media on UFO Investigations

As we have already discussed, the beginning of the modern UFO era began with the widely reported sighting by pilot Kenneth Arnold in 1947. Newspapers quickly seized on Arnold's description of "flying saucers," and the term became a household phrase overnight. The rapid dissemination of his account through print and radio set the stage for decades of public fascination.

Media coverage at this time served multiple functions. It informed the public, alerted authorities, and set the tone for public discourse. However, the limitations of early reporting, such as the lack of critical analysis and a tendency to sensationalize these encounters, also contributed to the first waves of UFO mania. Stories were often repeated with minimal verification, embedding myths deeply into the public consciousness.

Television Played a Key Role

The advent of television in the 1950s and 1960s marked a shift in how UFO stories were consumed. The visual medium allowed for greater dramatizations, re-enactments, and the presentation of photographic and video "evidence," which had a powerful impact on many viewers.

Televised reports could lend credibility to eyewitness testimonies or, conversely, expose hoaxes and inconsistencies. The coverage of iconic incidents like the 1961 abduction claim of Betty and Barney Hill, or the 1976 Tehran UFO encounter, engaged millions. Television's inherent dramatization often blurred the lines between fact and fiction, making it difficult for viewers to discern trustworthy information.

Documentaries, incident re-creations, and a large number of special broadcasts contributed to the mystique. They sometimes presented various degrees of alternative explanations and sometimes veered into speculative scientific and technical territory. Accuracy was not always guaranteed in these reports.

Public Pressure

Public pressure, fueled by media scrutiny, led to official investigations such as Project Blue Book in the United States. Periodically, declassified documents and official statements would receive coverage, sometimes quieting rumors and sometimes reigniting public debate. The cyclical nature of media attention created a feedback loop in which institutions felt compelled to respond, and their responses, in turn, created new waves of coverage and speculation.

Hollywood and the entertainment industry eagerly seized upon the UFO phenomenon. They enthusiastically produced feature films, television series, and many documentaries that both reflected and shaped public attitudes. Productions like "Close Encounters of the Third Kind" and "The X-Files" brought UFOs into mainstream culture, often blending scientific inquiry with elements of conspiracy and horror.

While these works contributed to a general fascination with the unknown, they also sometimes distorted public expectations. Fictional depictions blurred with reality, leading many people to accept dramatic narratives as being plausible explanations for real-life events.

The Advent of Digital Media and Its Capabilities

The rise of digital media, especially since the late twentieth century, further transformed the landscape of UFO investigations. The internet, social media, and smartphones made it possible for anyone to share sightings and theories instantaneously. This democratization has had both positive and negative consequences.

Greater visibility of sightings that might have once been ignored can now reach a global audience. Well-meaning citizens could use their own level of understanding for science to become amateur UFO investigators. When they would collaborate and jointly analyze their evidence, they would sometimes provide key insights that had been overlooked by many professionals. At other times, nonetheless, their curious conclusions were greatly opposed by actual scientists.

Disinformation has become a widespread fear. There were previously many instances of the unchecked spread of fake videos, digitally altered images, and unfounded conspiracy theories have been presented to the public. Artificial videos have become an even greater threat as digital manipulations have become more advanced and widespread. Due to this technology, seeing is not always leading to believing with imaging falsifications becoming so prevalent.

UFO Waves of Reports

Historically, spikes in mass media coverage have coincided with any waves of reported UFO sightings. When significant events or credible witnesses received coverage, the number of reports rises dramatically. Sociologists refer to this as the "contagion effect," where media attention acts as a catalyst for collective experiences, whether rooted in reality or merely in their perception.

Media Literacy and the Future of UFO Investigations

As public interest in UFO investigations remains high, the role of media literacy becomes assuredly more crucial. Audiences must learn to critically evaluate sources, try to distinguish between credible reporting and pure entertainment, and to maintain a healthy skepticism toward extraordinary claims. This has resulted in major educational initiatives, seeking ever more transparent governmental communications, and asking for a rejoinder from their responsible journalists. All of these have roles to play in ensuring that the exploration of the UFO phenomena should be guided by cautious reason and clear evidence rather than merely sensationalism and fear.

While we cautiously move forward, the challenge will be to harness the positive power of media. By doing so, we can promote open inquiry and informed debate, even as we guard against the pitfalls of sensationalism and misinformation. Only then can we hope to approach the enigma of UFOs with the clarity and rigor it deserves.

❖ The Galactic Impact of "Close Encounters of the Third Kind" on American Society

Few films have managed to leave as indelible a mark on the American psyche as Steven Spielberg's 1977 masterpiece, "Close Encounters of the Third Kind." Arriving at a time when the nation was emerging from the shadow of Watergate and grappling with questions about trust, the unknown, and humanity's place in the cosmos, this film struck like lightning.

It managed to capture our deepest imaginations and truly altered the country's cultural landscape. Its persisting influence has echoed for many years, across the film industry, the media forums, the scientific community, and within the collective consciousness. This unique motion picture was able to reshape the attitudes of many citizens around the world as they consider extraterrestrial life.

A Cinematic Turning Point

Before the release of "Close Encounters of the Third Kind," the portrayal of extraterrestrials in American cinema was largely characterized by fear, antagonism, and the specter of invasion. Alien films of the 1950s and

1960s, such as "The Day the Earth Stood Still" or "War of the Worlds," reflected both Cold War anxieties and a tendency to see the "others" as a threat to our own society.

Spielberg's film marked a true paradigm shift. Instead of depicting aliens as malevolent, it offered a vision of contact that was mysterious but ultimately benign. The movie was saturated with wonder and intense curiosity. This change would have far-reaching implications.

Influences on American Film and Storytelling

Spielberg's approach to considering extraterrestrial contact influenced an entire generation of filmmakers and storytellers. "Close Encounters" demonstrated that science fiction could be both highly imaginative and deeply personal, exploring intimate themes such as faith, communication, obsession, and reconciliation.

The film's protagonist, Roy Neary, was an every-person. He was just an ordinary American drawn into a fantastic cosmic mystery. This focus on the human experiences amidst such extraordinary circumstances opened the door for subsequent films. These would include "E.T. the Extra-Terrestrial," "Contact," and even "Arrival". It was now mainstream to blend speculative fiction with emotional depth.

The movie's technical achievements were plentiful. From Douglas Trumbull's mesmerizing visual effects to John Williams's iconic five-note musical motif, this film set new standards for the use of cinematic crafting. It proved that science fiction could garner critical as well as commercial success, paving the way for a whole new era of blockbuster filmmaking in America.

Impact on Science, Technology, and Education

The film's impact extended beyond art and into the sciences. "Close Encounters" fueled public fascination with astronomy, space exploration, and the search for extraterrestrial intelligence (SETI). Planetariums and science museums reported increased attendance, while school science programs used the film to spark interest in space and communications expertise. The movie's respectful treatment of scientific inquiry was embodied by the French researcher portrayed by François Truffaut. This character helped to legitimize the quest for understanding the universe in the eyes of the public.

Moreover, the film's theme of communicating with the unknown, by using music as a universal language, assuredly captured the imagination. It influenced many researchers in linguistics, semiotics, and the burgeoning field of computer science. The film inspired discussions about how humans might one day communicate with non-human intelligence, whether biological or artificial.

Influence on American Popular Culture

"Close Encounters of the Third Kind" quickly became a touchstone in American popular culture. Its imagery, from the glowing mothership to the sculpted mashed potatoes, entered the national lexicon. References to the film proliferated in television, advertising, and music, reflecting its deep integration into the American imagination. The five-note musical phrase became instantly recognizable, even to those who had not seen the movie.

Fostering a Sense of Wonder

Perhaps the film's most enduring legacy is the sense of wonder it instilled in its audiences. Spielberg's vision was one of awe in the face of mystery, urging viewers to approach the unknown with humility and hope. In a nation divided by politics and uncertainty, the film offered a unifying narrative. By utilizing the possibility that we are not alone, the movie expanded our expectations for our future encounters with aliens. This film gave us hope that contact with celestial beings could be guided by empathy and increased understanding.

The movie's conclusion, with its wordless exchange between the humans and the extraterrestrials, emphasized the power of creating connections beyond language and our differences. In doing so, it invited Americans to imagine a world in which curiosity triumphs over fear.

The final impact of "Close Encounters of the Third Kind" upon American and world-wide society is multifaceted and profound. It reshaped the portrayal of extraterrestrials in cinema, encouraged a spirit of scientific curiosity, and fostered a collective sense of awe. By challenging audiences to look beyond the familiar and to imagine the possibilities of contact with the unknown, Spielberg's film became more than just a cinematic milestone. It instead became a cultural event and a catalyst for conversation. It also became a beacon of hope in the age of skepticism.

We will never look at mashed potatoes the same!

❖ **The Ongoing Search for Government Transparency and Disclosure Related to UAPs**

Current calls for government transparency have grown louder, especially with recent admissions that some UAPs remain unexplained by science. In 2021, for example, the Office of the Director of National Intelligence released a report acknowledging over 140 incidents that basically could not be attributed to any form of conventional aircraft or atmospheric phenomena.

The roots of frustration remain at the heart of the UFO believers' desire for disclosure. Their annoyance lies with an enduring sense that critical information is being withheld by our own authorities. For many Americans, this is not merely conjecture. Instead, it is a core belief built upon decades filled with redacted documents, ambiguous statements, and unexplained official behavior.

The Role of Official Denial

Throughout history, the unique phenomenon of UFOs has absolutely captivated the imagination of many members of the public. This sparked great intrigue, engaged in speculation, and promoted myriad theories of extraterrestrial life. Despite the widespread fascination and countless reports of sightings, many governments worldwide have maintained an official position of denial or skepticism regarding the existence of UFOs as vehicles of alien origin.

Background of Governmental Evaluation of the UFO Phenomenon

Over the decades, several governments have conducted investigations into UFO sightings, often prompted by public pressure or national security concerns. Among the most notable are:

- The United States Air Force first established Project Sign in 1947. This was quickly followed by Project Grudge in 1949. The lack of success of these attempts led to the creation of Project Blue Book in 1952. This inquiry examined thousands of UFO reports and ultimately concluded, in 1969, that the majority of cases were primarily attributable to natural phenomena or man-made objects.

- The United Kingdom's Ministry of Defence (MoD), which undertook studies of UFO sightings following World War II. They lacked clear evidence, according to their released reports. Their officials declared that they found no evidence of threats or alien activity.

- The Soviet Union's investigations, during the Cold War era, were driven by concerns over potential foreign surveillance rather than extraterrestrial involvement. The details of how far they went in their investigations did not become clear until after the eventual fall of the Soviet Union. Similar to those officials in the Western governments, they failed to reveal any evidence of alien origins of the UFOs they evaluated.

The Governments' Possible Reasons for Continued Official Denial

The reluctance of world governments' decisions to deny any form of acknowledgement for the existence of UFOs as actual extraterrestrial phenomena can be attributed to several factors. Nevertheless, the main factor has been viewed as the most significant, and that is thought to be because of the perceived need for assuring national security. Denials by officials supposedly helps prevent public panic and ensures that sensitive defense mechanisms remain confidential.

Is this really required by the general public, or is it time we were finally told the truth?

The Sustained Lack of Conclusive Evidence

Despite numerous reports and speculations, there has been limited concrete evidence to prove that UFOs are truly of alien origin. Governments tend to base their conclusions on scientific analyses, based on revealed information. Anything to avoid admitting that UFOs are real!

Mechanisms of official denial are quite similar around the planet. Bureaucratic denunciations often take the form of carefully worded statements, explaining nothing. They use the declassification of reports with inconclusive findings to avoid controversy. They utilize redirection of public inquiries to ensure that uncertain scientific explanations abound.

Governments have frequently employed the use of secrecy and subjective classification efforts to manage uncomfortable information about the reality of UFOs. Based on what we have seen so often, if your governmental agency does not like the answers developed to these awkward questions, it is easier to simply lie. Then, lie again. After that, lie again. Lie as often is needed!

Public Skepticism and Conspiracy Theories

The lack of governmental acknowledgment has led many to believe in cover-ups and hidden truths about extraterrestrial involvement. Conspiracy theories, ranging from secret alien treaties to hidden technologies, have proliferated as a result. This is a reasonable reaction when the public has heard of evidence that has been denied the light of day. Trust in governmental institutions greatly suffers when their citizens do not believe what they are being told by their political leaders!

The Public's Consistent Demand for Disclosure

In recent years, the ongoing demand for governmental disclosure regarding UFOs has evolved into a global movement. Advocates argue that transparency is not just a matter of intellectual curiosity, but also of public interest and trust. Incidents involving airline pilots, military personnel, and even commercial satellite data suggest that unidentified aerial phenomena could have implications for both safety and security. But are the details of the truth being released?

We don't believe so!

The release of the U.S. government's 2021 report on unidentified aerial phenomena was hailed as a milestone by some, but many UFO believers found it rather disappointing. While the report did acknowledge that dozens of incidents remained unexplained, it offered little in the way of concrete evidence or any new revelations. For many in the UFO community, this was yet another example of incremental changes in transparency that ultimately falls short of true disclosure.

The Internet Age: Amplifying Frustration

The rise of the internet and social media has both fueled and complicated the drive for transparency. On one hand, whistleblowers and independent researchers have had more tools available at their disposal than ever before. On the other side, the proliferation of misinformation and hoaxes makes it easier for skeptics to dismiss the subject entirely.

This strange digital cacophony can intensify the frustrations of genuine believers, who must now fight not only for access to hidden truths but also for the credibility of reports in a very crowded and often sensationalized field.

The Broader Implications of Official Secrecy

For many UFO believers, the lack of governmental transparency has implications that go beyond the subject of extraterrestrial life. The withholding of information is seen as an affront to democratic principles and the public's right to know. This tension underscores a much wider debate about the balance between national security and open governance.

Moving Forward with Great Hope and Stifling Challenges

Despite their frustrations, UFO believers continue to persevere in their quest for information. Many remain hopeful that future disclosures will finally provide the answers they seek. However, they also recognize the formidable challenges that lie ahead.

These include recognized bureaucratic inertia and extends to secret corridors of power. The difficulty of comparing interests of national security and public disclosure are real. Until the barriers of secrecy are lowered and the full scope of what is known is brought to light, the frustration is likely to persist.

Still, optimistic hope springs eternally that disclosure will happen!

Here Is an Important Detail from My Past

I will make this point very clearly: I have PTSD

Post-Traumatic Stress Disorder is a term that sometimes still carries a stigma associated with it. Too many people think this means a person is weak, unstable, or even disabled. As with many medical conditions, one key element is how severely a person is affected by the cause of their disorder.

While I will be using the term "soldiers" for this section, this condition is exactly the same for sailors, Marines, and aerial combatants as well. No person is truly immune to the effects of severe trauma upon our minds!

Many folks consider the concept of psychological trauma resulting from the horrors of war as a modern phenomenon, closely associated with the twentieth century and the twenty-first century. It brings to mind conflicts such as World Wars I and II, the Korean War, Vietnam, the Gulf Wars, Afghanistan, and so on. Yet the personal injury to veterans of these ongoing struggles needs to be viewed against the backdrop of historical perspectives.

You see, the seeds of war-induced psychological distress have existed since humans first took on combat. Recently discovered archaeological documentation has revealed that what we consider to be PTSD today was fully recognized, even among the Spartans. These legendary, brave warriors of olden Greece were fully susceptible to the effects of this condition. By using historical records, modern psychological research, and the timeless human experience of trauma, we can finally understand this condition far better.

Current Treatment Methods for Soldiers with PTSD

Using evidence-based medical approaches and emerging therapies, we have developed fresh ideas on how we should approach individuals suffering with this malady. Post-Traumatic Stress Disorder is a mental health condition that can occur following exposure to traumatic events such as combat, severe violence, or true disasters. Soldiers, however, are thrust into their unique exposures to intense combat and life-threatening situations by the very nature of their service. For this reason, they are particularly at high risk for developing PTSD.

Understanding PTSD in Soldiers

PTSD is characterized by symptoms such as intrusive memories, nightmares, flashbacks, avoidance of reminders of trauma, negative changes in mood and cognition, and heightened mental arousal. The experience of PTSD can have profound effects on a soldier's personal and professional life, impacting relationships, careers, and their overall well-being. I can assure you that PTSD also affects all those around them, for their families and close friends will commonly suffer as well.

Early recognition and intervention are crucial for effective treatment. The stigma and career concerns about this diagnosis often deter soldiers from seeking help, highlighting the need for supportive, confidential, and accessible treatment environments.

Psychotherapeutic Treatments

Psychotherapy remains the cornerstone of PTSD treatment, with several modalities demonstrating efficacy, particularly for military populations. The American government's Veterans Administration has recognized this and placed increased emphasis in recent years. I receive this treatment from the VA myself.

Cognitive Behavioral Therapy (CBT)

CBT is widely regarded as the gold standard for PTSD treatment. It focuses on changing unhelpful patterns of thinking, behavior, and emotional responses related to trauma. Several specialized forms of CBT are used for treating PTSD in soldiers:

- Cognitive Processing Therapy (CPT): CPT helps individuals process and reframe negative beliefs about trauma. Through structured sessions, soldiers learn to challenge and modify unhelpful thoughts that contribute to PTSD symptoms.

- Prolonged Exposure (PE) Therapy: PE involves gradually and safely confronting trauma-related memories, feelings, and situations that have often been shunned in their past. This process reduces fear and avoidance and helps individuals regain control over their responses to any trauma reminders.

- Eye Movement Desensitization and Reprocessing (EMDR)

  EMDR is an evidence-based therapy that uses guided eye movements to help individuals process and integrate traumatic memories. Research indicates EMDR can be particularly effective for military personnel who may struggle with verbalizing traumatic events. This is particularly significant because of how so many individuals endure this condition in relative silence.

- Group Therapy

Group therapy provides a supportive environment where soldiers can share experiences, learn coping strategies, and decrease their feelings of isolation. Group sessions may be structured around trauma processing, skills training, or peer support. Being able to communicate with those who have had similar experiences can be highly important, since many victims do not feel comfortable in discussing combat events to non-combatants. Combat survival guilt, which is all too common, can be especially improved in this fashion.

- Family and Couples Therapy

As mentioned above, PTSD often affects not only the individual but also their family and friends. Family and couples therapy can often improve the lines of communication, reduce conflict tension, and foster understanding for why the victim behaves in ways they would not normally do. It is a joint effort to help families cope with the impact of PTSD for those they love.

- Pharmacological Treatments

Medications can play an important role in managing PTSD symptoms, particularly when combined with psychotherapy.

o Selective Serotonin Reuptake Inhibitors (SSRIs), such as sertraline, fluoxetine, and paroxetine, are frequently the first-line pharmacological treatments for PTSD. These medications increase the levels of serotonin within the brain.

This is a necessary neurotransmitter, so if these levels are low, it can be important to improve their relative levels within crucial regions of the brain. They are effective in reducing core symptoms like anxiety, depression, and intrusive thoughts. These medications are generally well-tolerated and have a favorable safety profile.

It is not uncommon, from what I have witnessed in my own patients, that these cannot always be viewed as a group. One of these medications may have no effect on an individual, while a second agent could produce unwanted side effects. However, trying the third medication may be highly beneficial to that same patient. The very medication that one patient hates can be one that another patient absolutely loves. These are extremely personalized responses by way of treatment.

o Serotonin-Norepinephrine Reuptake Inhibitors (SNRIs)

SNRIs, such as venlafaxine, may be prescribed if SSRIs are ineffective. They target similar serotonin neurotransmitter systems but also affects norepinephrine levels as well. These medications can often help with both mood and anxiety symptoms. Once more, it may be necessary for a patient to try a dose, then possibly try an even higher dose before deciding that this medicine is not right for them.

o Prazosin

Prazosin is an alpha-1 adrenergic antagonist that was originally used for blood pressure control. As is often the case, doctors and patients discovered additional benefits while using this medication. The later use of that medication may be completely different from its original purpose. Such is the case here. Prazosin has shown remarkable efficacy in reducing nightmares and improving sleep in many soldiers with PTSD. It is often used as an adjunct to other treatments.

o Other Medications

Benzodiazepines are sometimes helpful, but are generally avoided due to their addictive potential and often limited efficacy. They can, however, sometimes assist with persistent anxiety.

Mood stabilizers, antipsychotics, and other antidepressants may be considered in complex cases or when there are co-occurring conditions such as clinical depression or bipolar disorder. Both of these latter conditions will usually require their own treatments and patient understanding in order to comprehensively improve the lives of these victims.

• Complementary and Alternative Therapies

An increasing number of soldiers are turning to complementary and alternative approaches to support their recovery. While the evidence base is still growing, these methods can be useful adjuncts to standard care. As with the belief in UAPs/UFOs, the attitude someone carries into the doctor-patient conversations can also improve or limit their acceptance of these therapies.

- Mindfulness, Meditation, and Prayer

Mindfulness-based interventions, such as mindfulness-based stress reduction (MBSR) and unique meditations, have been shown to decrease PTSD symptoms, reduce stress, and improve overall well-being. I am including prayer in this context because I choose to believe in God. I believe in the power of prayer, so it is a basic conviction for me personally. These practices encourage a focus upon present-moment awareness and self-compassion.

- Yoga and Physical Activity

Yoga, tai chi, and other forms of gentle physical activity can alleviate anxiety, improve mood, and promote relaxation. Structured programs tailored to veterans may be especially beneficial in specific cases.

- Acupuncture

Some studies suggest acupuncture may help reduce symptoms of PTSD, particularly anxiety and insomnia. While more research is needed, acupuncture is considered safe and may be offered as a complementary therapy.

- Innovative and Emerging Treatments

As our understanding of PTSD and its associated symptoms deepens, new and innovative treatments are constantly being explored, and there can be breakthroughs at any time.

- Virtual Reality Exposure Therapy (VRET)

VRET uses immersive virtual environments to safely expose soldiers to trauma-related cues in a controlled setting. This interesting technology can be particularly effective for those who struggle with traditional exposure therapy or who have difficulties in accessing in-person care.

- Stellate Ganglion Block (SGB)

SGB is a minimally invasive procedure involving the injection of a local anesthetic near the stellate ganglion nerve cluster. By blocking the neural pathways in this cluster, it appears to be specifically effective in some people. Early studies suggest SGB may rapidly reduce PTSD symptoms in these individuals, though more research is needed.

- MDMA-Assisted Psychotherapy

Clinical trials are investigating the use of MDMA (commonly known as ecstasy) or marijuana in conjunction with psychotherapy for severe, treatment-resistant PTSD. Early findings are promising, with some soldiers experiencing significant symptom reductions. However, these treatments remain entirely investigational and are not yet widely available. This is a very controversial topic today.

- Transcranial Magnetic Stimulation (TMS)

  TMS is a non-invasive neuromodulation technique that uses magnetic fields to stimulate specific brain regions associated with mood and trauma processing. TMS has shown potential as an adjunctive treatment for PTSD, particularly for those who do not respond to other therapies.

  As we learn more about the electromagnetic neural effects in our own brains, we are expanding our understanding of just how complex we are as humans. This is an area of possible crossover effects that can be used with witnesses to UAP events.

- Comprehensive and Integrated Care

  Effective treatment for soldiers with PTSD often involves a holistic, integrated approach. Many military and veteran health systems offer multidisciplinary care teams composed of psychiatrists, psychologists, social workers, occupational therapists, and peer specialists. These teams work collaboratively to address the complex and interconnected needs of soldiers, including physical health, substance abuse, housing, and vocational supports.

- Peer Support Programs

  As has been mentioned before, peer support programs can be extremely helpful. The ability to speak in a protected environment is powerful. The understanding of fellow veterans who have lived with analogous experiences leading to their PTSD can truly assist fellow soldiers, can reduce their perceived stigma. It can also foster greater hope for progress. It may promote the victims' own engagement in cooperating with other forms of care they might otherwise neglect. Being able to discuss topics and events they normally avoid can be very therapeutic in these settings.

Hearing about the success of other veterans in treating this condition is also a great way to encourage the success for each individual listening.

- Telehealth and Digital Health Solutions

Telehealth platforms expand access to care, particularly for soldiers in remote or underserved areas. Digital health tools, such as mobile apps for symptom tracking and self-management, may very well supplement in-person treatment and provide ongoing accessible support to make a real difference for those who might not receive care otherwise.

- Barriers to Care and Future Directions

Despite the availability of effective treatments, many soldiers do not seek or receive timely care due to factors such as guilt, the perceived stigma of being labelled with this term. There might also be a sad lack of awareness or limited access to care.

Many victims have substantial concerns about the confidentiality of what they have felt and experienced. It requires continued efforts to educate military communities, to reduce the PTSD stigma, and to expand access to evidence-based and emerging treatments.

Research into the biological underpinnings of PTSD, personalized medication approaches, and novel therapeutics continues to evolve the treatment for this condition. The key integrations of other technologies, such as artificial intelligence for symptom monitoring and personalized treatment recommendations, may further enhance the effectiveness for the care of these soldiers.

Conclusion

The treatment landscape for soldiers with this condition is promising. Global collaboration and technological advancements may soon yield deeper empathy and understanding for patients, their families, and even health care providers. Those with PTSD may require care that is dynamic and multifaceted. It could include encompassing psychotherapies, medications, complementary approaches, and cutting-edge innovations.

Through individualized, evidence-based, and compassionate care, there is great hope for helping these soldiers. Treatments can assist them to recover, reclaim their lives, and thrive beyond their past trauma. Ongoing research, advocacy, and support remains vital to ensuring that all soldiers have access to the help they need.

We Got it From Them!

Chapter 9

The Combined History and Future of UAP Research

Interest in UAPs is growing, with renewed legitimacy among scientists. New organizations, such as the Galileo Project at Harvard, aim to apply rigorous methods to the study of UAPs, using advanced sensors and open data sharing. This is a private attempt to increase our understanding of what is happening within our cosmos. We will specifically address this later on.

Understanding U.S. Government Responses to Unidentified Aerial Phenomena

The subject of unidentified aerial phenomena has long intrigued both the public and authorities in the United States. In recent years, increasing transparency and a surge in reported sightings have prompted new interest in UAPs from the U.S. military agencies, the Pentagon, and the Federal Aviation Administration. I want to explore in detail how these institutions are changing the manner by which they review and respond to new reports of UAPs, the processes involved, and the implications for national security and public understanding.

The Pentagon's Role in UFO Investigations Has Been Dubious

The Pentagon, as the headquarters of the U.S. Department of Defense, plays a central role in UAP investigations. The DoD involvement has grown significantly in the past decade, largely due to increased reporting by military personnel and the potential implications for airspace security. AARO, the All-Domain Anomaly Resolution Office, is the current Pentagon agency making an attempt to provide a significant response to military UAP reports. Let's look further into this process.

The UAP Task Force

o   Before AARO was launched, the Pentagon developed the UAP Task Force that was responsible for investigating military encounters with UAPs. Established in 2020, the Task Force supposedly focused on standardizing data collection and analysis across the branches of the armed forces. Because of the perceived laissez-faire attitude they used, this program was discontinued in favor of the initiation of AARO.

In 2022, the Department of Defense established their All-Domain Anomaly Resolution Office, as we have described. Officially, AARO has been tasked with affirmed responsibilities that include:

o   Collecting and analyzing data from military sensors and eyewitnesses
o   Coordinating with intelligence and civilian agencies
o   Assessing risks to military personnel and flight operations
o   Reporting their findings to Congress and other oversight bodies
o   They provide authoritative reviews and reports that are submitted by military pilots, radar operators, and other personnel.

These reports are analyzed for their potential to represent advanced technology, foreign surveillance, or unusual natural phenomena. The office uses a variety of analytical tools, including radar data, infrared imagery, and pilot testimony to make their assessments.

The Military Review Process for New UAP Reports

When a military pilot or radar operator observes a UAP, the following steps are typically taken:

- Initial Detection: The object is spotted visually or picked up on radar or other sensors.

- Immediate Reporting: Military protocol dictates that unusual or unidentified aerial activity must be rapidly reported through the appropriate chains of command. Standardized forms are used to capture the details.

- Data Preservation: Radar tracks, flight data, video, and audio recordings are preserved for further analysis.

- Preliminary Assessment: The unit's intelligence officers conduct an initial assessment, looking for explanations such as equipment malfunction, weather, or known aircraft.

- Escalation to Higher Authority: If no explanation is found, the case is escalated to the service branch's headquarters and then to the Pentagon's AARO reporting system.

- Further Analysis: AARO, or the relevant Pentagon offices, need to bring together experts in the fields of radar, advanced optics, cutting-edge aerospace engineering, and the bureaus of intelligence to thoroughly review each case.

  It is questionable for how much faith the public still has in this necessary function to be performed in an unbiased fashion.

Classification and Reporting

Most UAP reports from military personnel are initially classified, as they involve sensitive details about detection capabilities and military operations. However, the Pentagon periodically releases unclassified summaries to the public and to Congress.

The FAA's Role in UAP Reports

The Federal Aviation Administration is directly responsible for the safety of civil aviation within U.S. airspace. It routinely receives reports of strange aerial phenomena from commercial pilots, air traffic controllers, and occasionally from members of the public.

FAA Reporting Mechanism

When a commercial pilot or air traffic controller observes a UAP, the incident is reported through established communication channels. The FAA collects information such as:

- Date, time, and location of the sighting

- Altitude, speed, and direction of the observed object

- Weather conditions at the time of observation

- Sensor data (e.g., radar tracks)

- Eyewitness accounts

Coordination with the Pentagon and NORAD

If the FAA determines that a UAP sighting could pose a risk to national security or air safety, it coordinates with the North American Aerospace Defense Command, more commonly known as NORAD. Referencing other sources within the Pentagon, suitable FAA reports are forwarded to AARO for further analysis.

Public Accessibility and Data Sharing

While the FAA keeps records of UAP reports, it often refers inquiries to the military or to the National UFO Reporting Center (NUFORC) for civilian sightings. Data sharing between the FAA, the Pentagon, and civilian agencies has improved in recent years, though much information remains classified as sensitive.

Government Transparency and Congressional Oversight

In response to public demand for transparency, Congress has required regular reporting from the Pentagon's UAP investigating offices. Major actions include:

- Public release of declassified UAP videos and reports

- Congressional hearings featuring testimony from military officials and UAP witnesses

- Mandates for annual or semi-annual reports to congressional intelligence and armed services committees

These measures aim to ensure that the government investigates UAPs thoroughly while keeping the public informed about potential risks and discoveries.

Recent High-Profile UFO Reports

Several recent incidents have fueled public and governmental interest:

- USS Princeton and USS Nimitz Encounters in 2004: As covered before, Navy aviators reported and recorded multiple UAPs off the coast of California. The objects displayed flight characteristics that could not be easily explained. Besides that, there were many other UAPs repeatedly identified in the area.

- East Coast Navy Encounters from 2014 to 2015: F/A-18 carrier pilots and support vessels detected fast-moving UAPs on radar and visually, sparking a wave of internal reporting. Once more, there were "swarms" of UAPs that were recurrently noted to be flying in the vicinity of the carrier strike group.

- The 2023 Congressional Hearings: Several former military personnel testified that UAPs represented a national security concern and called for increased transparency. Even more detailed interviews were evaluated by Congress in 2024.

The Challenges in Reviewing UFO Reports

Despite advances in sensor technology and analytical methods, reviewing UAP reports poses several challenges:

- Data Quality: Many reports lack high-resolution imagery or comprehensive radar data. Nevertheless, many other reports demonstrated remarkable clarity for their events.

- Sensitive Technology: Revealing methods of UAP detection could compromise national defense. This would cause these events to be withheld due to the classification of key incident details.

- Stigma and Underreporting: Pilots and controllers may still hesitate to come forward with their complete stories for fear of ridicule or adverse career repercussions. Back in 1992, this certainly affected how I dealt with what I had witnessed.

- Ambiguity of Reported Phenomena: Natural atmospheric events, man-made objects, and technical glitches can mimic UAPs. Going back to Project Blue Book, these types of simple explanations often strike UAP enthusiasts as being used by the government to further their attempts to cover-up the true sources for these investigations.

Ongoing Research and Future Developments

The government continues to refine its approach to UAPs. Research initiatives seek to:

- Enhance detection and sensor fusion capabilities

- Develop databases for standardized reporting

- Encourage reporting by removing the stigma, especially for military personnel

- Expand interagency cooperation, to include the Pentagon, NORAD, the FAA, and even the FBI. There is a need to have multi-disciplinary evaluations and event-sharing capabilities promoted more eagerly across governmental agencies.

The goal is not only to safeguard airspaces but also to advance scientific understanding of these unexplained phenomena.

Conclusion

The review and investigation of new UAP reports by the U.S. military, the Pentagon, and the FAA is already a complex and ever-evolving process. Analyses can be driven by the dual imperatives of assuring national security while ensuring proper scientific curiosity.

With a range of institutional reforms, dedicated offices, and increasing transparency, the U.S. government continues their open attempts to build a more systematic and credible approach to understanding unidentified aerial phenomena. Whether these objects represent advanced foreign technology, natural phenomena, or something yet unknown, thorough review processes ensure that every credible sighting is carefully evaluated, and the public remains informed about this enduring mystery of the skies.

❖ The Galileo Project at Harvard

The Galileo Project at Harvard is a groundbreaking initiative that seeks to definitively answer one of humanity's most profound questions: Are we alone in the universe?

Named after Galileo Galilei, the pioneer of modern science who famously challenged established paradigms about our cosmos, the project aims to bring the search for extraterrestrial intelligence into mainstream scientific inquiry. Spearheaded by Professor Avi Loeb of Harvard University, the Galileo Project is an ambitious SETI endeavor to detect, analyze, and understand any extraterrestrial technological artifacts that may exist within or beyond Earth's atmosphere.

Overview of the Galileo Project

The Galileo Project was launched by Harvard in 2021. This decision followed widespread interest in the extraterrestrial phenomena. This was particularly in light of the U.S. government's release of declassified reports on UAPs. These reports highlighted the existence of aerial objects observed by military pilots that exhibited characteristics defying known human technologies. While the origins of these phenomena remain unexplained, they underscore the necessity of a systematic, scientific approach to investigating potential non-human intelligence.

The primary goal of the Galileo Project is to establish a robust framework for the detection and analysis of extraterrestrial technological signatures. Unlike traditional SETI, which focuses on detecting signals such as radio waves from distant civilizations, the Galileo Project emphasizes the search for physical artifacts, potentially even alien technologies, that can be seen within Earth's vicinity and beyond.

The Vision and Objectives

The Galileo Project operates on the principle that scientific rigor, transparency, and open inquiry should guide the search for extraterrestrial intelligence. Its objectives can be outlined as follows:

1. Detection of UAPs

The project aims to employ advanced observational tools, including telescopes, radar systems, and high-resolution cameras, to detect and track Unidentified Aerial Phenomena. By creating a global network of observation sites equipped with cutting-edge technology, the Galileo Project hopes to gather empirical data on UAPs and identify their origins.

2. Search for Extraterrestrial Artifacts

The Galileo Project seeks to locate and analyze any potential extraterrestrial artifacts on Earth or within our solar system. This includes objects that may have originated from interstellar sources, such as "Oumuamua." This enigmatic interstellar object was detected in 2017, which exhibited unusual characteristics that prompted wide-ranging speculation about its nature.

3. Advancement of SETI Techniques

By integrating data-driven methodologies, artificial intelligence, and machine learning algorithms, the project aims to refine traditional SETI approaches and develop innovative new techniques for identifying any form of extraterrestrial evidence.

4. Public Engagement and Education

The Galileo Project is committed to fostering public interest and understanding of extraterrestrial research. Outreach efforts such as lectures, publications, and collaborations with other educational institutions are integral to its mission.

Technological Framework

The Galileo Project relies on cutting-edge tools to achieve its ambitious goals. Some of the key components include:

- Telescopic Arrays: A far-ranging network of telescopes equipped with high-resolution imaging capabilities to observe UAPs and celestial objects.

- AI and Machine Learning: Algorithms have been designed to analyze vast amounts of observational data and to detect even trace patterns of evidence that could possibly be indicative of alien technology.

- Interdisciplinary Collaboration: Partnerships with a variety of physicists, engineers, astronomers, and other experts. This type of collaboration is needed to ensure a comprehensive approach to the study of evidence for any extraterrestrial phenomena.

Scientific Controversy and Challenges

The Galileo Project has sparked debates within the scientific community. While many praise its innovative approach, others question the feasibility of detecting extraterrestrial artifacts. Many scientists argue that the focus on UAPs may detract from more traditional astrophysical research.

Professor Avi Loeb is the Frank B. Baird Jr. Professor of Science at Harvard University. He has addressed these concerns by emphasizing the need for rigorous science and highlighting the project's potential to expand our understanding of the cosmos.

Funding and resource allocations pose additional challenges to this controversial project. As the project primarily relies on private donations and corporate grants, sustaining long-term operations naturally requires significant ongoing financial support.

Breakthroughs and Future Prospects

Although the Galileo Project is still in its early stages, it has already achieved several milestones. The establishment of distributed locations for observational facilities and the initiation of data collection mark significant progress in the search for extraterrestrial intelligence. Future plans include expanding to even more observational networks, collaborating with international research institutions, and jointly publishing their findings to share insights within the global scientific community.

Conclusion

The Galileo Project at Harvard represents a bold step forward in the quest to understand our place in the universe. By combining scientific rigor with visionary ambition, this private project not only challenges conventional boundaries but also inspires humanity to look beyond Earth for answers to age-old questions. As it continues to evolve, the Galileo Project has the potential to unlock mysteries that could redefine our understanding of life and intelligence in the cosmos.

❖ **The FAA's Plans to Investigate Future UAP Incidents**

An Overview of Emerging Protocols and Implications

The Federal Aviation Administration is now poised to take on a more proactive role in investigating UAPs. This decision comes amidst growing public interest and government acknowledgment of unexplained aerial phenomena that have been noticed within American airspace. With heightened attention from media, researchers, and official policymakers, the FAA's move signals a significant step towards addressing the mysterious sightings that have captured so many imaginations for multiple decades.

Context and Background

UAPs have long been a subject of public fascination and concern. Reports of strange, unidentified objects in the skies have persisted globally. This has sparked debates about their origins and their potential implications. Historically, UAP investigations were relegated to organizations like the military or the intelligence agencies. However, the FAA is responsible for ensuring aviation safety and efficiency for the flying public. It seems it is now entering this arena with their key focus being upon safeguarding our airspace and enhancing governmental transparency.

The Role of the FAA

As the primary regulatory body for civil aviation in the United States, the FAA has a critical responsibility to maintain safe skies. While safety protocols traditionally focus on addressing risks such as technical failures or human error, the inclusion of UAP investigations represents a paradigm shift for them. The agency aims to analyze these incidents within the basic framework of airspace management and to understand any potential safety risks that could be posed by UAPs.

New Protocols for Investigating UAP Incidents

The FAA plans to establish standardized procedures for handling UAP reports. These protocols will enable pilots, air traffic controllers, and other aviation professionals to document and report their observations in a systematic manner. Key components of these protocols include:

- Incident Reporting Systems: A streamlined process for civilian pilots and commercial aviation personnel to file their detailed accounts of UAP sightings. This will be complete with both visual and radar data.

- Data Analysis Framework: Collaboration with advanced scientific institutions and serious researchers to analyze the collected data rigorously, ensuring its objectivity and its credibility.

- Coordination with National Defense: Measures will be taken to coordinate with the military and with intelligence agencies to address any potential security threats associated with UAPs. With the military having difficulties in the past assuring truly unbiased reporting, this is likely to remain a difficult challenge into the future.

Technological Implications

The FAA's approach to investigating UAPs will likely leverage advanced technologies, including satellites, radar systems, and even artificial intelligence. These ever-improving tools are essential for capturing high-resolution images, tracking flight paths, and identifying factual anomalies. The hope is that enhanced surveillance capabilities will provide a more comprehensive understanding of UAP behaviors and their flight characteristics.

Collaborations With Scientific Communities

To ensure robust investigations, the FAA intends to partner with universities, research institutions, and independent experts. By fostering collaboration, the agency seeks to develop a multidisciplinary approach that combines aerodynamics, astrophysics, and aviation expertise. This cooperation will also help dispel myths and misinformation surrounding UAP sightings.

Enhanced Safety Measures

The FAA's investigation of UAP incidents will aim to strengthen aviation safety standards by identifying and mitigating risks associated with unidentified aerial phenomena. For instance, there is a strong need to understand how UAPs interact with aircraft and airspace. Better awareness of these characteristics could lead to improved safety protocols for avoiding mid-air collisions.

Transparency and Public Trust

A significant benefit of the FAA's involvement in UAP studies is the desire for a potential increase in case transparency. By addressing these phenomena openly, the FAA can foster greater confidence among the public and aviation professionals. Detailed reports and thoroughly researched findings may also reduce many speculative theories and therefore provide much clearer insights into the very nature of UAP incidents.

Shaping Policy

The FAA's investigations could influence broader government policies related to aerial phenomena. Findings may necessitate revisions to international airspace agreements, defense protocols, and aviation safety measures. Additionally, the FAA's efforts could pave the way for global cooperation on UAP research.

Challenges Ahead

While the FAA's plans are ambitious, several challenges must be addressed:

- Resource Allocation: Investigating UAPs require significant levels of funding and may require specialized personnel. These factors might strain the FAA's existing financial resources.

- Scientific Rigor: Ensuring that investigations are grounded in scientific evidence is crucial to avoid sensationalism. In addressing this, the FAA will need to cautiously determine which scientists would be asked to contribute to these evaluations.

- Interagency Collaboration: Coordinating with military and intelligence agencies may present bureaucratic hurdles. After all, the Pentagon and these intelligence agencies do not typically open their arms to outside investigations.

Conclusion

The FAA's decision to investigate future UFO incidents marks a bold and pivotal moment in the study of aerial phenomena. Through systematic protocols, technological advancements, and key collaborations, the agency aims to enhance flight safety, ensure transparency, and to increase understanding of these incidents. While challenges remain, the FAA's involvement underscores the importance of addressing unexplained phenomena in a scientifically rigorous and responsible manner. As humanity continues to explore the skies and beyond, the FAA's efforts may ultimately lead to profound discoveries about our universe and our place within it.

What about Britain?

❖ British Governmental Plans to Investigate UFOs

An Exploration of Their Stated Policies and Initiatives

The study of UAPs has captivated public imagination in Britain for decades, blending elements of scientific curiosity with a widespread fascination for the unknown. While UAP investigations have historically been led by private organizations and enthusiastic individuals, governmental attention to the phenomenon has gained momentum in various nations. This includes the United Kingdom. The country has a storied history of addressing reports of aerial anomalies, and recent initiatives suggest an evolving approach to UAP investigations.

Historical Context

In the years immediately following World War II, sightings of UAPs began to garner widespread attention. The British government, like many others, became interested in these reports, primarily from a defense perspective. Concerns about unidentified aircraft violating airspace led the Ministry of Defence, MoD, to quietly monitor such occurrences. The "Flying Saucer Working Party" was created by the MoD in 1950. This was Britain's first official attempt to analyze UFO phenomena. This group concluded that sightings were likely attributable to natural phenomena or misidentified aircraft, and its investigations were largely discontinued.

The Ministry of Defence's UAP Desk

From the 1960s to 2009, the MoD operated a dedicated UAP desk that was tasked with collecting, analyzing, and documenting reports from the public and from military personnel. The desk's primary function was to assess whether any sightings posed a threat to national security. Over the years, numerous reports were filed, ranging from peculiar lights in the sky to detailed accounts of unknown craft. While the desk did not find any evidence supporting extraterrestrial vehicles, its existence underscored governmental interest in maintaining UAP oversight.

This is where Nick Pope comes in.

As a civilian agent, Pope worked for the Ministry of Defence from 1985 to 2006. From 1991 to 1994, he reportedly worked in the Secretariat of the Air Staff. His location and duties were commonly known as the so-called "UFO desk." As such, his duties included examining the reports of UFO sightings within British territory. Meanwhile, the Ministry of Defence officials specified that the government would remain open-minded about their research.

They at last concluded that UFO sightings did not pose any threat to the UK, and that no findings of extraterrestrial craft had been found. Pope, nonetheless, wrote that he did not share the MoD's final opinion. Since he retired, he has become an ardent UAP enthusiast.

Recent Developments

As global interest in UAPs reignited during the 2020s, spurred in part by similar investigations in the United States, the British government faced vocal calls to revisit its position on UAPs. Parliamentary debates occasionally touched on the subject, with some lawmakers urging for more transparency and an updated investigative framework. Public petitions demanded that the government disclose historical UAP files and implement more robust inquiry mechanisms that have also gained traction.

Advances in Technology and Data Collection

Modern technology has transformed the possibilities for UAP investigations. Having enhanced radar systems, imaging devices, and artificial intelligence offers unprecedented capabilities for detecting and analyzing anomalous aerial phenomena. Recognizing these advancements, British defense agencies have expressed interest in leveraging these updated technologies to improve their monitoring of the skies. This is particularly important as threats from drones and other unmanned systems continue to evolve.

Collaboration with International Allies

The United Kingdom has historically collaborated with its allies on matters of defense and intelligence, and UAP investigation is no exception. Given the United States' establishment of the All-domain Anomaly Resolution Office to systematically study UAPs, Britain may consider forming a similar entity or to simply align its efforts with the American initiatives.

Scientific Inquiry

There is growing advocacy for scientific engagement in UAP research. Many experts suggest that investigations should focus on collecting empirical data and analyzing phenomena under strict scientific protocols. Universities and research institutions could play a pivotal role in such endeavors, potentially receiving funding for projects related to aerial anomalies.

Public Skepticism

Despite renewed British governmental interest, public skepticism about UAP investigations remains quite significant. Some people view such inquiries as distractions from more pressing issues, while others suspect that the government may never fully disclose its true findings about UAPs. It will never be able to fully satisfy all of its citizens.

Conclusion

While the British government's plans to investigate UAPs are still emerging, they reflect broader global trends where nations are increasingly taking the phenomenon seriously. Whether fueled by concerns about national security or genuine curiosity about the unknown, these efforts mark an important step toward understanding what lies beyond our usual perception of the skies. As science and policy converge, the future of UAP investigations in Britain holds promise, intrigue, and the potential for groundbreaking discoveries.

❖ **South American Aspects of the UAP Problem**

The various governments in South America have faced similar problems to those elsewhere around the planet. There are also large numbers of famous UAP incidents the have happened in those nations. As such, South America has long been a hotspot for UAP sightings, with numerous incidents that have captivated the imaginations of believers and skeptics alike

The region's vast and diverse landscapes, from dense jungles to high mountain peaks to sprawling deserts, truly provide ample opportunities for mysterious phenomena to emerge. Let us examine some of the most famous UAP incidents in South America, shedding light on stories that have intrigued the world.

The Varginha Incident, Brazil

One of the most famous UAP cases in South America is the Varginha Incident, which occurred in January of 1996 in the state of Minas Gerais, Brazil. Local residents reported seeing a strange craft crash near the town. Shortly afterward, several witnesses claimed to have seen strange humanoid beings with brown, oily skin and large red eyes wandering in the area. These beings were described as having a peculiar odor and moving in an unusual manner.

The Brazilian military and police were allegedly involved in capturing one or more of these beings, leading to widespread speculation about their own government cover-ups. While official reports denied the existence of any form of extraterrestrial life, the incident remains a cornerstone of UAP lore within Brazil.

The Bariloche Airport Incident, Argentina

In July of 1995, an airline pilot flying into San Carlos de Bariloche Airport in Argentina reported a close encounter with a UAP. The pilot's account described a glowing object that appeared to move at incredible speeds and perform maneuvers that were impossible for conventional aircraft.

The incident gained credibility when airport personnel corroborated the pilot's testimony, stating that they had observed a similar object on their radar. The encounter actually disrupted air traffic temporarily, and the event remains one of Argentina's most compelling UAP sightings.

The Chulucanas Lights, Peru

Peru has also had its share of UAP sightings, with one of the most well-known being the strange Chulucanas Lights. In October of 2001, residents of Chulucanas, a small town near Piura, witnessed strange lights in the sky that danced and changed colors. The phenomenon was reportedly witnessed by hundreds of locals over several nights.

The lights were captured on video, sparking global interest and debate. While some skeptics attributed the phenomenon to natural atmospheric events, believers pointed to the consistency and unusual behavior of the lights as evidence of extraterrestrial activity.

The Colares UAP Attacks, Brazil

Another chilling UAP case from Brazil is the Colares UAP attacks, which took place in the late 1970s in the Amazonian town of Colares. Residents reported being attacked by beams of light emitted by unidentified flying objects. These beams allegedly caused burns, dizziness, and other physical symptoms to the local inhabitants who witnessed these bizarre craft.

The Brazilian Air Force then launched an investigation, dubbed Operation Saucer, to study the phenomena. Although the military collected testimonies and photographs of UAPs, the official report ultimately dismissed the claims, attributing them to mass hysteria. Nonetheless, the incident continues to fuel speculation and intrigue.

The Petare UAP, Venezuela

In the 1950s, the Petare neighborhood of Caracas, Venezuela, became the site of a famous UAP sighting. Witnesses reported seeing a large, saucer-shaped object hovering in the sky. Some claimed that the craft remained stationary for several minutes before abruptly disappearing while traveling at incredible speeds.

This sighting was notable for its number of witnesses and the clarity of their descriptions. It became one of Venezuela's earliest and most well-documented UAP incidents, sparking widespread interest in extraterrestrial phenomena.

The Nazca Lines and Theories of Ancient Alien Contact

While not a modern UAP incident per se, the Nazca Lines in Peru are often linked to extraterrestrial theories. These massive geoglyphs, created by the Nazca culture somewhere between 500 BCE and 500 CE, clearly depict animals, plants, and geometric shapes. Some theorists suggest that the lines were created as landing strips for UAPs or as signals for alien visitors.

Though archaeologists attribute the lines to ceremonial and astronomical purposes, their sheer scale continues to inspire speculation about their connection to beings from other worlds. For an ancient people to devote the time and efforts these geoglyphs would require, it remains fascinating to imagine why this culture felt the need to create them all.

South America's Enduring UAP Legacy

South America's UAP phenomena extends beyond isolated incidents, encompassing a broader cultural fascination with the unknown. The region hosts numerous conferences and gatherings for UAP enthusiasts, where researchers and witnesses share their findings and theories. Documentaries, books, and films have also kept these stories alive, ensuring that the mysteries of the skies remain firmly embedded in the public consciousness.

Whether driven by genuine encounters, natural phenomena, or merely human imagination, South America's UAP incidents continue to captivate and inspire. As technology advances and our own understanding of the universe deepens, perhaps one day we will uncover the truths behind these enigmatic sightings.

❖ African nations have their own UAP problems!

Africa is a continent rich in culture, history, and natural beauty. Nonetheless, it has also been home to some intriguing UAP incidents that continue to spark curiosity and debate. These mysterious encounters have fascinated researchers, skeptics, and enthusiasts alike, leaving behind questions that remain unanswered to this day.

The Ariel School Incident of 1994

One of the most well-documented UAP incidents in Africa took place on September 16, 1994, at the Ariel School near Ruwa, Zimbabwe. On this day, over sixty schoolchildren reported seeing strange beings and an unidentified flying craft that landed in the school yard during their morning break. The children described the awesome beings as having large heads and big black eyes. Some of these children even claimed to have received telepathic messages from them.

What makes this case particularly compelling is the consistency of the children's accounts, their emotional reactions, and the lack of evidence suggesting they were being influenced externally. Investigators, including renowned psychiatrist Dr. John Mack, interviewed the children and found their stories coherent and convincing. The Ariel School Incident remains one of the most intriguing UAP encounters globally, largely due to the sheer number of witnesses and the detailed descriptions they provided of these surprising events.

The Credo Mutwa Accounts

Credo Mutwa, who was a renowned Zulu shaman and cultural historian, has often spoken about his encounters with unidentified flying objects and extraterrestrial beings. Mutwa claimed that such sightings were not new to African history and lore, suggesting that UAP incidents are woven into the spiritual and mythological tapestry of the different cultures across the continent.

One of Mutwa's most striking accounts involves a craft he encountered during one of his spiritual journeys. While some skeptics view Mutwa's narratives as symbolic or allegorical, others believe his stories may point to a deeper connection between African spirituality and their beliefs in the truth of extraterrestrial phenomena.

The Lesotho UAP Crash of 1995

In 1995, reports emerged about a possible UAP crash in Lesotho. This is a completely landlocked country surrounded by the territory of South Africa. According to witnesses, a strange object was seen hurtling through the sky before crashing in a remote area. Local villagers reported seeing unusual lights and hearing loud noises during the incident. These people believed this was an utterly unique event.

While investigations into the crash yielded no definitive evidence, the story gained traction among UAP enthusiasts worldwide. Some believe the object was surely an extraterrestrial craft. Many others argue against this presumption. They have attributed the event to be just falling space debris or some sort of a misidentified natural phenomenon.

The Kalahari Desert Incident of 1989

The Kalahari Desert, sprawling across the countries of Botswana, Namibia, and South Africa, was the setting for a controversial UAP incident in May of 1989. According to reports, a military aircraft allegedly intercepted and engaged with a UAP, which then crashed in a remote desert area. There were witnesses who claimed to have seen military personnel quickly recovering the craft and even possibly extraterrestrial bodies.

The South African government denied any knowledge of the event, leading to speculation and a number of conspiracy theories. Some argue the incident was nothing but a hoax, while others continue to investigate the possibility of actual extraterrestrial involvement. The witnesses of the area certainly believed in the latter explanation.

Other UAP Sightings Across Africa

Africa has witnessed numerous other UAP sightings, with many remaining unexplained to this day. From strange lights observed over the skies of Nigeria to peculiar formations of craft seen in Kenya and Ethiopia, these occurrences continue to fuel the debate about extraterrestrial activity on the continent.

- In Uganda, residents near Lake Victoria reported seeing luminous objects hovering over the water, sparking theories about underwater alien bases.

- In Egypt, mysterious lights near the pyramids have been linked to UAP activity, blending ancient history with modern mysteries.

- South Africa, particularly the Cape Town area, has been a hotspot for UAP sightings, with reports of glowing objects zigzagging across the sky.

The Cultural Significance of UAPs in Africa

The vast array of UAP incidents across Africa often intertwine their tales with the continent's rich cultural and spiritual beliefs. Indigenous communities sometimes interpret these odd sightings as manifestations of ancestral spirits or divine beings. This certainly adds layers of meaning to the phenomena. This cultural context provides unique insights into how UAPs are perceived in Africa compared to other parts of the world.

Challenges in Investigating UAP Incidents

Investigating UAP incidents in Africa comes with its own set of challenges. Remote locations, limited access to scientific equipment, and cultural barriers often hinder comprehensive studies. Additionally, skepticism from authorities and the general public can make it difficult to pursue serious research into these unusual occurrences.

The South African UAP Wave of the 1970s

South Africa, as one of the continent's most developed countries, has seen a large number of UAP sightings reported over time. In particular, the 1970s witnessed a massive wave of sightings across the country. These were unidentified objects primarily described as luminous disks or cigar-shaped crafts appearing in the night sky.

These accounts, so similar to those reported on other continents, often came from rural areas, where witnesses were able to observe the phenomena away from urban light pollution. Some reports were accompanied by sincere claims of electromagnetic interference with electronic devices, adding to the mystique.

With advancements in technology and growing interest in UAP phenomena, Africa may play a pivotal role in uncovering new evidence about extraterrestrial life. Organizations and independent researchers are beginning to collaborate to document and analyze sightings across the continent. As Africa's skies continue to be a source of intrigue, the possibility of groundbreaking discoveries remains tantalizingly within reach.

Africa's UAP incidents, steeped in mystery and cultural significance, offer a fascinating glimpse into the unknown. From the chilling accounts of schoolchildren in Zimbabwe to the enigmatic crash in Lesotho, these stories challenge us to expand our understanding of the world. At the same time, it could grant us insights into the universe around us. Whether one believes in extraterrestrial life or views such incidents through the lens of skepticism, the allure of Africa's skies remains undeniable.

❖ Famous UAP Incidents in Asia

Let's begin a journey into a multitude of unexplained phenomena.

The mysterious realm of UAPs have captivated humanity for decades, and Asia is no exception. Across the continent, tales of strange lights, otherworldly vehicles, and inexplicable occurrences have intrigued and puzzled both skeptics and believers alike. Let us delve into some of the most famous UAP incidents in Asia, exploring their fascinating narratives and the enduring questions they leave behind.

The Kofu Incident in Japan in 1975

One of Japan's most well-known UAP encounters occurred on February 23, 1975, in the city of Kofu, in the Yamanashi Prefecture. Two young boys reported seeing a metallic, saucer-shaped object land in a vineyard. According to their testimony, strange humanoid figures emerged from the craft. The boys described them as wearing silver suits and having large ears. The boys then claimed that the beings attempted to communicate with them before returning to their spacecraft and departing. The incident was investigated extensively, yet no definitive explanation was found, leaving the Kofu Incident as one of Japan's enduring UAP mysteries.

The Delhi Flying Object of 1964

In January of 1964, residents of Delhi, India's capital, witnessed an extraordinary event. Hundreds of people reported seeing a bright, glowing object hovering in the sky, and then moving erratically before disappearing. The UAP was described as disc-shaped and emitted a peculiar hum.

The incident drew widespread attention, and authorities were inundated with the number of witness accounts they received. Despite an array of speculation, ranging from weather balloons to true extraterrestrial visitation, no conclusive determination was ever made. With the cultural influences of Indian religious beliefs, as we have discussed earlier, this incident was taken very seriously.

The Phoenix Mountain Incident of China in 1994

China's Phoenix Mountain Incident is one of the most detailed UAP cases in the country. On October 9, 1994, Meng Zhaoguo, a farmer in the

Heilongjiang Province, claimed to have seen a glowing object that descended onto the Phoenix Mountain. He further alleged that he had physical contact with extraterrestrial beings and described intricate details of their appearance and their technology.

The case was subjected to scientific investigation, including polygraph tests of witnesses, but opinions remain divided on its authenticity. The Phoenix Mountain Incident continues to be a topic of debate among China's UAP researchers and enthusiasts.

The Malaysia Airlines MH370 Mystery

While not officially categorized as a UAP incident, the disappearance of Malaysia Airlines Flight MH370 in 2014 captivated the world's imagination and spurred theories involving extraterrestrial activity. The Boeing 777 disappeared without a trace, sparking extensive search operations and speculation about the role of unidentified aerial phenomena. It is possible that one of the pilots intentionally flew the airliner off into the Indian Ocean. While no evidence has been discovered to substantiate either of such claims, the case remains one of aviation's greatest mysteries.

Recurring Sightings Around Mount Fuji

Mount Fuji, Japan's iconic peak, has long been associated with UAP sightings. Numerous hikers, tourists, and locals have reported seeing luminous objects hovering or darting around the mountain. These sightings are often accompanied by intriguing photographs and videos, though skeptics point to natural phenomena like lenticular clouds or misidentified aircraft as potential rationalizations. This is an explanation that is not commonly believed by the witnesses of these UAPs. Regardless, Mount Fuji continues to be a Japanese hotspot for UAP enthusiasts.

The Manila UFO Sightings in the Philippines in 2011

UAPs have long captured the imagination of people around the world, including those in Manila, the bustling capital of the Philippines. Over the decades, various accounts of mysterious lights, strange shapes in the sky, and unexplained phenomena have emerged, adding Manila to the global tapestry of UAP lore.

Early accounts were quite interesting. The fascination with UAPs in Manila dates back to the mid-20th century. This was when Philippine newspapers occasionally reported strange aerial occurrences. Witnesses described certain erratic movements of strange luminous objects that defied conventional explanations like airplanes or weather balloons. While concrete evidence remained elusive, such repeated reports stirred a great deal of curiosity among locals and researchers alike.

In recent years, the advent of smartphones and social media has led to an increase in UAP-related claims from Manila residents. Videos showing glowing lights and peculiar movements in the skies have gone viral, sparking debates among both skeptics and believers. Some sightings were attributed to drone displays, atmospheric phenomena, or optical illusions, while others distinctly defied any easy categorization for these events. There are many UAP believers in the Philippines and they are not easily dissuaded from their opinions.

Among the notable instances, one event stands out: a reported sighting in the late evening above Quezon City, in the Metro of Manila. Witnesses described a cluster of lights moving in irregular patterns. This created a buzz among UAP enthusiasts nationwide. While some experts speculated that it was a rare meteorological event, others pointed to extraterrestrial activities instead as a possibility for the origin of these lights.

Scientists and skeptics in the Philippines often approach UAP sightings with caution, advocating for rigorous investigation before jumping to conclusions. Most phenomena are explained by natural or man-made causes, such as the normal culprits: reflections, aircraft, or weather conditions. Nevertheless, the mystery surrounding the remaining cases lack definitive answers. Because of this, the event has assuredly fueled ongoing interest in the UAP area of research.

Cultural Impact

UFO sightings have also found their way into Philippine pop culture, inspiring movies, television shows, and books that explore the idea of extraterrestrial contact. For many, these stories symbolize the boundless curiosity and imagination of humanity, transcending scientific explanation to delve into the unknown. Some people even harken to some of their ancient beliefs in their arguments about the validity of these events.

While Manila may not be known as a global hotspot for UFO activity, the city's stories and repeated sightings contribute to the broader tapestry of unexplained phenomena. Whether viewed as evidence of extraterrestrial life or simply misinterpreted occurrences, their fascination with UAPs remains a testament to humankind's enduring curiosity about the universe.

The Cultural Impact of UAP Sightings in Asia

The recurring UAP sightings within Asia have deeply influenced many local cultures. These sightings have inspired art, literature, and especially modern media. From traditional folklore in India blending extraterrestrial themes to modern cinematic portrayals of alien encounters, the obvious fascination with UAPs transcends generations. In countries like Japan and China, dedicated UFO researchers and reporting societies have even emerged, fostering a community of enthusiasts and investigators.

Conclusion

Asia's famous sets of UAP incidents continue to fuel international curiosity and wonder. While definitive answers may remain elusive here, just as it does even in the United States, the many stories serve as serious reminders of humanity's boundless imagination and its desire to explore the unknown.

There is a long history of religious beliefs concerning past encounters with gods that came from the sky. In Asia, these beliefs transcend generations and remain a large part of the cultures of these age-old civilizations.

❖ **The Australian Government's Investigation of UFOs**

The modern discourse on unidentified flying objects has taken a significant leap in recent years as governments worldwide are increasingly acknowledging their importance. These events are not accepted merely as science fiction but as legitimate subjects of study. Aside from basic curiosity, these sightings occur with potential implications for national security, aviation, and science. Australia, known for its vast skies and open landscapes, has joined the ranks of countries looking to better understand the enigma of UAPs.

Background on UAP Investigations

The study of UAPs on this distant region has historically been shaped by international efforts, including Australia's steps to integrate its own approach. Combined with the long-held mysteries associated with the famed Outback, and from its numerous other curious geological features, this continent is certainly involved with a persisting history of UAP events.

Australia's Historical Engagement with UAP Phenomena

While the Australian government has not maintained a high-profile program akin to the U.S. Department of Defense's investigations, Australia has a storied history of UAP sightings. As has been noted previously, UAPs are not simply an American phenomenon.

The Notorious Westall UAP Incident of 1966

The Westall UAP incident remains one of Australia's most compelling and mysterious UAP stories. On April 6, 1966, over 200 students, teachers, and local residents of the Melbourne suburb of Clayton South, claimed to have witnessed an unidentified flying object in broad daylight near the Westall High School.

What followed was a flurry of curious intrigue, probing government involvement, and decades of speculation. These elements have kept this case alive in the annals of UAP history.

The Events of April 6, 1966

The day began like any other at the Westall High School, but it quickly turned into something extraordinary. At approximately 11 AM, students and teachers noticed a strange, silvery-gray object in the sky. Witnesses described the object as disk-shaped, about the size of two cars, and moving silently. It hovered and maneuvered with remarkable agility, unlike conventional aircraft. The craft fulfilled the definition of a UFO.

The object reportedly descended and landed briefly in a nearby open field known as The Grange, a paddock surrounded by trees. Excited and curious, many students ran toward the area to get a closer look. Some claimed that the object left behind a circular marking on the grass before taking off again at high speed, pursued by several smaller aircraft, believed by some to be military planes.

Witness testimonies from that day paint a vivid yet varied picture of the event. Many students recall a feeling of awe and fear, while teachers appeared more skeptical. Some witnesses claimed to have seen the object emit a faint glow, while others noticed no unusual lighting. What remains consistent, however, is the sheer number of people who saw something unusual that day.

News of the sighting spread quickly, attracting media attention from local newspapers and radio stations. However, there was no official explanation from government authorities. Witnesses reported that shortly after the incident, individuals in military uniforms arrived at the scene, cordoned off the area, and allegedly warned people not to discuss what they had seen. Some claimed that school officials were also instructed to keep the matter quiet.

Over the years, the absolute lack of transparent communications from governmental officials has fueled theories about a cover-up. Was it an experimental aircraft? A weather balloon? Or was it something not of this world? The absence of clear answers has left the incident shrouded in mystery.

The Westall UFO incident has sparked numerous theories over the decades:

1. Military Experiment

One popular theory suggests that the unidentified object was a secret military aircraft being tested by the Australian or U.S. governments. Advocates of this idea point to the alleged presence of military personnel at the scene and the subsequent secrecy surrounding the event.

2. Natural Phenomenon

Skeptics propose that the sighting could have been some type of natural phenomenon, such as an unusual cloud formation. However, there were many witnesses who absolutely disputed this explanation. They have cited the object's speed, maneuverability, and peculiar shape as being inconsistent with any known natural or manmade phenomena.

3. Extraterrestrial Origin

For UAP enthusiasts, the Westall incident represents compelling evidence of extraterrestrial visitation. The object's behavior, combined with the number of witnesses and the lack of a plausible alternative explanation, has made this case a cornerstone of Australia's UAP lore.

Impact and Legacy

The Westall UAP incident has become a cultural touchstone in Australia and beyond. It has inspired documentaries, books, and even commemorative events. In 2013, a plaque and park were dedicated to the incident, ensuring that the lasting memory of this mysterious event will remain alive for future generations to ponder.

For the witnesses, the event was life-changing. Many have spoken about how the experience altered their perception of reality. Meanwhile, others have expressed frustration at the lack of recognition or investigation by authorities. Despite varying interpretations of what happened that day, the Westall incident has left an indelible mark on those who experienced it.

The Australian Westall UAP incident of 1966 is a fascinating story of mystery, intrigue, and human curiosity. Over 50 years later, it remains unsolved, captivating the imagination of skeptics and believers alike. Whether it was a secret military project, a natural anomaly, or an encounter with the unknown cannot be proven.

The truth of what happened that day continues to elude us. We can be certain that the Westall UFO incident will remain as one of Australia's most enduring mysteries.

The Kaikoura Lights of New Zealand in 1978

Though not geographically part of Australia, the close proximity of New Zealand causes me to add this story here. The Kaikoura Lights incident is often discussed in the context of regional UAP phenomena due to its proximity to both Australia and Asia.

On December 21, 1978, strange lights were seen by the crew of a cargo plane flying over Kaikoura in New Zealand. The objects were not only visually noted, but were also tracked on radar and appeared to follow the flightpath of the involved aircraft. Investigators reviewed radar data and witness accounts, but the lights remain unexplained to this day, becoming a landmark case in UAP history.

The Royal Australian Air Force (RAAF) has previously been tasked with handling reports of UAPs. However, its role in investigating these claims has been minimal and largely procedural, often classified under broader aviation safety measures. In recent years, as scientific and public interest surged, calls have grown louder for a more structured approach to studying the phenomena. The government is reportedly considering even more in the way of systematic efforts to investigate UAPs. While details for studies in this field remain under wraps, discussions appear to center on the following key objectives:

1.   Collaboration with International Partners

Given the global nature of UAP sightings and investigations, Australia seeks to collaborate closely with other nations. They will likely join with allies like the U.S. and the U.K. Both countries have established protocols and ongoing research initiatives. Combined efforts might include sharing data, jointly developing common methodologies, and even operational strategies for tracking and analyzing these sightings.

2. Scientific Research and Technological Integration

The government aims to leverage Australia's strong scientific community, including organizations like the Commonwealth Scientific and Industrial Research Organisation (CSIRO), to develop technologies and strategies for monitoring and understanding UAPs. This approach emphasizes the use of radar, satellite imaging, and artificial intelligence to analyze patterns and anomalies within the continent's airspace activity.

3. Public Transparency and Reporting Mechanisms

An essential element of the governmental plan is fostering public trust. Establishing easy-to-use platforms for civilians to report sightings, coupled with commitments to transparency in findings, will likely be key to this initiative. This aligns with international trends, where governments have recognized that open communication can enhance citizen engagement and reduce unfounded speculation.

4. Ensuring Aviation Safety and National Security

One of the primary drivers behind governmental interest in UAPs is the potential impact on aviation safety and national security. This is true for nearly all governments, and it remains true for Australia as well. Unidentified objects in civilian airspace could pose risks to commercial and military aircraft alike.

By understanding their nature, they seek to determine whether these UAPs represent advanced foreign technology, natural phenomena, or something altogether unknown. The government can better ensure the safety of its skies if it could more definitively determine how to deal with these UAPs.

This seems to be a task no government can handle.

Potential Challenges

Despite the promising trajectory of the plans, there are several hurdles Australia may face in its pursuit of UAP investigations. They are largely the same obstacles present for the U.S. Congress as they follow similar lines of inquiry.

- Resource Allocation: Developing an investigative framework that involves technological integration and collaboration may require substantial funding and manpower. Within budget constraints, where does the money come from to pursue these studies?
- Public Skepticism: UAP research often carries the stigma of a pseudoscience, and educating the public on the seriousness of investigations will be crucial. In current discussions, this will continue to be difficult for both proponents and skeptics to agree upon. In the rowdy politics of recent years, it is difficult enough to even pass a national budget, without tossing in so great of hindrance as UAP skepticism.
- Interagency Coordination: A cohesive effort will require collaboration among various governmental and scientific agencies. These measures may initially face significant bureaucratic hurdles.

Determining the line between national security and the public's interest to know what is happening blurs the arguments when dealing with the military.

Australia's decision to delve deeper into the study of UFOs reflects a broader societal shift. As humanity becomes increasingly interconnected through global challenges. Competing with topics like climate change and space exploration, the investigations of UAPs represents another frontier where science, technology, and security intersect. Moreover, this initiative could inspire educational programs, scientific curiosity, and international cooperation.

Public Reaction and Expectations

The Australian public has shown mixed reactions to the government's emerging interest in UFO investigations. Enthusiasts and researchers have welcomed the move, seeing it as validation of years of anecdotal evidence and grassroots documentation. Skeptics, however, question the allocation of resources to an area they perceive as speculative. The government will need to balance these perspectives by presenting a clear, evidence-based rationale for its actions.

As Australia joins with the United States by stepping into the realm of UFO investigations, it joins a growing list of nations seeking clarity on these phenomena that have long fascinated humankind. Whether these efforts will lead to groundbreaking discoveries or merely confirm mundane explanations remains to be seen.

What is certain, however, is that Australia's skies, so vast and enigmatic, hold the promise of unveiling mysteries that could reshape our grasp of the world. Perhaps they could even help us to more easily understand our place in the universe.

❖ American and Other Citizens-Based Scientific Observations and Reporting Agencies

Amateur astronomers and everyday American people contribute valuable data through organized observation networks and reporting platforms. Citizen science may prove crucial in distinguishing true anomalies from noise. Here is some history about this:

- The National Investigations Committee on Aerial Phenomena (NICAP): It was founded in 1956 by inventor Thomas Townsend Brown. NICAP quickly became one of the most influential civilian UAP research organizations of its day. Comprised of scientists, military veterans, and laypeople alike, NICAP advocated for transparency in government dealings with UAP data and conducted independent investigations of sightings. Its eventual demise in 1980 led other organizations to rise and take the lead in UAP evaluations.

- The Mutual UFO Network (MUFON): Established in 1969, MUFON remains one of the world's largest and most enduring civilian UFO research groups. With thousands of volunteers, MUFON collects reports, conducts field investigations, and offers a training program for investigators. MUFON's global presence allows for the aggregation of data from a variety of cultures and regions.

- British UFO Research Association (BUFORA): Across the Atlantic, BUFORA has played a similar role in the United Kingdom since its formation in 1962. It focuses on the scientific investigation of UFOs and provides public education on aerial phenomena. The BUFORA Journal was produced from 1962. For 43 years, it was the mainstay of the association's contact for its members and enthusiasts. However, the website fully replaced the magazine in 2005. It carries on events today.

- Centro de Estudios de Fenómenos Aéreos Inusuales (CEFAI) and Other International Groups: Many countries have their own citizen-led UFO organizations, such as CEFAI in Argentina, GEPAN in France (though GEPAN is affiliated with the French space agency CNES), and numerous regional groups throughout Europe, Asia, and Latin America.

The International UFO Bureau was established in Oklahoma City in 1957 by a young and passionate ufologist named Hayden C. Hewes. Hewes had a degree in aeronautical and space engineering. He was inspired by the growing number of UFO sightings and the public's insatiable curiosity during the Cold War era. This was a time fraught with both technological optimism and existential unease. Hewes sought to create a dedicated hub for the collection, investigation, and dissemination of UFO reports.

At the time, Oklahoma, like much of the American Southwest, was experiencing its own share of skyward mysteries. Reports of strange lights, disc-shaped craft, and unexplained phenomena were regularly circulating in newspapers and whispered about in diners and living rooms. Hewes recognized the need for a formal bureau that would not only catalog these reports but also attempt to analyze them objectively. His bureau was formed to be free from government secrecy or ridicule. It was one of the earliest privately run, citizen-based UFO organizations in the U.S.

The IUFOB's mission was threefold:

- To collect and investigate reports of unidentified flying objects from around the world

- To provide a supportive environment for those who had experienced UFO encounters or sightings

- To promote open-minded scientific inquiry into the UFO phenomenon

The Bureau quickly gained national attention. Hewes, a tireless self-promoter, launched newsletters, sent press releases, and invited local and national media to cover the Bureau's activities. He had a radio show that had hundreds of thousands of devoted followers. He encouraged correspondents from across the United States, and even internationally, to submit their UFO experiences by mail, phone, or in person. Each report was carefully compiled, soundly analyzed, and often followed up with his radio interviews. He also led site visits and answered requests for physical evidence, such as photographs or sketches.

For many, the IUFOB became a first point of contact. It was a place where witnessers could share their extraordinary experiences without the fear of mockery. The Bureau's files soon swelled with reports, ranging from the commonplace to the sensational. These included accounts of close encounters, alleged abductions, and even purported government cover-ups.

Notable Cases and Contributions

The Bureau also became an early advocate for the scientific investigation of UFOs. Hewes and his team corresponded with other prominent ufologists, including his work with Dr. J. Allen Hynek and with Jacques Vallée. He sought to bridge the gap between the public's fascination and the scientific establishment's skepticism. The IUFOB held seminars, panel discussions, and even organized sky-watching events designed to educate and engage the community.

To connect with both enthusiasts and the merely curious, the International UFO Bureau published a range of newsletters and pamphlets. The most prominent of these was the "UFO Bureau Bulletin," a regular publication that featured case studies, eyewitness accounts, editorials, and updates on ongoing investigations. The bulletin was distributed nationally and, at its peak, reached numerous thousands of loyal subscribers.

The Bureau also maintained an extensive library of books, periodicals, and files related to the UFO phenomenon. This resource was open to the public, offering researchers, journalists, and students access to one of the largest privately held UFO archives in the region.

Outreach extended beyond just the printed page. Hewes became a sought-after guest on radio and television shows, appearing on local and national programs to discuss the Bureau's work and the broader implications of UFO sightings. The IUFOB's open-door policy fostered a sense of inclusivity, attracting people from all walks of life, to include farmers, police officers, teachers, and even skeptics. All were welcomed into these conversations.

In 1973, just as the future President Jimmy Carter's national profile grew, he decided to officially document a UFO event he had personally witnessed. He eventually chose to file his report with the International UFO Bureau in Oklahoma City. Carter filled out the standard forms as requested, describing the date, time, and circumstances of the sighting in as much detail as he could recall. A copy of this report is available in his presidential library.

Key details from Carter's report include:

- Date of sighting: 1969 (some later reports mistakenly cite 1973, the date the report was filed rather than the sighting itself).

- Estimated time: 7:15 PM.

- Location: Leary, Georgia.

- Duration: 10-12 minutes.

- Number of witnesses: Around ten.

- His Description: Bright white object, changed color to blue and red, no sound, hovered and then moved away.

Carter was clear in emphasizing that he did not believe the object was an alien spacecraft, but rather something he simply could not identify. Whether this was a true "unidentified flying object," the report was one of many filed during the period. However, its significance was amplified by Carter's subsequent rise to the presidency.

Nevertheless, the legacy of the International UFO Bureau endures and is now growing once more. It pioneered many investigative techniques that were later adopted by larger organizations. It has fostered open dialogue around anomalous phenomena, and, most importantly, provided a community for those who had experienced the unexplained.

The IUFOB's files are bulging, in terms of computer memory. They remain a valuable historical record, capturing a unique era in American culture when the sky truly seemed full of mystery.

The remarkable story of the International UFO Bureau is emblematic of the broader interest in unidentified flying objects that swept across America in the latter half of the 20th century. Oklahoma, with its wide horizons and open spirit, proved fertile ground for such inquiry. The IUFOB's influence can be seen in today's events and the public's ongoing fascination with UFOs. It promotes key documentaries and a wide array of podcasts to renew government interest in UAPs.

In the recent days, under the strong guidance of CEO Mindy Tautfest and the IUFOB's Board of Directors (one of whom is ME!), the story of the IUFOB is not just about past unidentified flying objects. It is more about the unending human desire to seek answers to the profoundest questions, to wonder, and to never stop looking up at the sky in search of something more. By adding more and more serious investigators, and by adding vital leaders with former military and key technical expertise, the Bureau certainly sees its brightest days ahead.

- Project Hessdalen: In Norway, a long-standing partnership between amateur researchers, local citizens, and academic leaders has led to continuous monitoring of anomalous lights in the Hessdalen valley. As a result, they have produced several scientific papers and feature real-time data streams.

Additionally, various citizen-based groups have begun to employ more advanced technologies:

- Automated Sky Surveillance: Networks of cameras, radar, and motion sensors that will continuously scan the skies for any anomalous events. These techniques can generate vast amounts of data for analysis.

- Data Analytics: Machine learning and pattern recognition algorithms are now being leveraged to sift through massive databases in search of correlations and trends. Additional AI teams could drive great technological advances in this field.

- Citizen-Led Projects: Open-source projects like SETI@home (originally for the Search for Extraterrestrial Intelligence) have inspired similar distributed computing efforts related to UFO data analysis. People can use their own computers and volunteered time to further these investigations.

Collaboration with Official Agencies

Although early relations between citizen groups and official agencies were sometimes adversarial, often characterized by mistrust and mutual suspicion, in recent years there have been a great deal more cooperation. Governments in the U.S., U.K., and elsewhere have declassified thousands of UAP files, often at the urging of citizen activists. In some countries, authorities now accept reports directly from the public, recognizing the value of grassroots observation networks.

The Sociocultural Impact of UAP Citizen Groups

Beyond the search for extraterrestrial life, citizen UAP groups serve important cultural functions. They operate as community forums for people to share extraordinary experiences without fear of ridicule, fostering connection and discussion. In some cases, the collective investigation of unexplained phenomena has led to the discovery of new meteorological, astronomical, or psychological processes.

Moreover, these organizations have been instrumental in shaping popular narratives around UFOs, influencing everything from literature and film to government policy. By democratizing the investigation of the unknown, citizen groups have helped maintain public interest and a spirit of inquiry.

Despite their achievements, citizen-based UAP groups face persistent criticisms:

- Lack of Scientific Rigor: Critics argue that anecdotal evidence and subjective reporting hinder objective analysis.

- Fragmentation: The proliferation of small, uncoordinated groups can lead to duplication of effort and the utilization of inconsistent standards of veracity.

- Credibility: The presence of hoaxes, conspiracy theories, and blatant sensationalism within some communities has damaged the public perception of these UAP research groups.

However, many organizations have responded by professionalizing their procedures, partnering with academics, and promoting healthy skepticism alongside open-mindedness.

The Future of Citizen UFO Studies

The landscape of UFO research is changing rapidly. Especially as governments begin to take the subject more seriously. This is evidenced by the formation of the U.S. Department of Defense's All-domain Anomaly Resolution Office and increased congressional attention. Despite all of this, citizen groups continue to play a vital role. Their grassroots efforts ensure that UAP phenomena can remain a matter of public record and scrutiny.

Citizen groups (such as the International UFO Bureau) and individual researchers have been at the heart of UFO studies for over seventy years. Their key contributions range from data collection and field investigation to fostering increased public discourse. They have kept the mystery of UFOs alive in the public imagination. In a world increasingly shaped by technology and information, their quest endures: to shine light on the unexplained, and perhaps, one day, to unravel one of humanity's most captivating enigmas.

Conclusion

The enigma of UFOs endures at the frontier of science and human curiosity. While most sightings have earthly explanations, a persistent core of cases challenges conventional understanding, inviting open-minded yet skeptical inquiry. Whether future discoveries reveal new natural phenomena, advanced technology, or even distant civilizations, the scientific study of UFOs serves as a compelling reminder: the universe still holds many secrets, waiting to be explored by those willing to look up and wonder.

❖ The Future of the American Congressional Hearings on UFOs

Transparency and Accountability Have Been Promised, Yet We Still Search for Answers

In recent years, congressional hearings on UAPs have intently captivated the American public and sparked a renewed interest in the mysteries that linger within our skies. Once relegated to the realm of pop culture, conspiracy theories, and late-night talk shows, UAPs are now the subject of serious governmental inquiry.

The U.S. Congress has been holding hearings, demanded increased transparency, and has pressed for answers from military and intelligence communities. The future of these important proceedings promises to reshape the way society understands this enigmatic phenomenon.

Historical Congressional Context

The U.S. government's engagement with UAPs stretches back to the 1940s and 1950s, with programs like Project Sign, Project Grudge, Project Blue Book, and a few congressional hearing. Congress more recently request the Department of Defense to create the Advanced Aerospace Threat Identification Program (AATIP), followed by the current AARO.

For decades, information on UAPs was tightly controlled and often shrouded in official secrecy. This, conversely, did not silence those who wanted the government to disclose more information. The lack of open transparency instead led to widespread speculation and distrust.

However, a series of leaked videos in the late 2010s depicting UAPs encountered by naval aviators, was coupled with official acknowledgment of their validity from the Department of Defense. These revelations set the stage for a new era of public discourse.

The first modern congressional hearings on UAPs occurred in the 1960s, but it wasn't until May 2022 and July 2023 that the House of Representatives held open hearings to address UAPs. Lawmakers pressed for answers on unexplained sightings, the possible existence of non-human technology, and whether there are government programs operating without oversight.

The Drivers Behind Renewed Congressional Interest

Several factors have converged to make UAPs a topic of legitimate governmental concern:

- National Security: Unidentified aerial objects may pose a threat to military and civilian airspace. In particular, "drones of unknown origin" have repeatedly invaded the restricted airspace above Langley Air Force Base. These unidentified flying objects did this with complete impunity.

  If foreign adversaries possess this type of technology that the United States cannot identify, it raises serious defense concerns. If they are instead craft created by non-human intelligences, it raises even more serious defense concerns.

  Can we really control the skies above our own airbases? It doesn't appear to be the case!

- Transparency and Accountability: Members of Congress and the public have grown wary of perceived secrecy within intelligence agencies. Lawmakers demand clear information, oversight, and the assurance that no clandestine programs are running unchecked. President Trump has been a mover and shaker for release of the Kennedy assassination files and many researchers seek this same type of transparency for UAPs.

- Public Pressure: Public fascination with UAPs remains very high. Witness testimonies from credible military personnel, pilots, and intelligence officials have lent legitimacy to the subject. By doing so, these testimonies have fueled greater demands for more open dialogue.

- The public fallout from the release of the Navy's Tic Tac, Gimble, and GoFast videos has been intense. The public wants to see and hear more of any similar videos hiding within the vaults of the Pentagon. The DoD surely has more than the three videos that have been released to this point.
Show us more videos!

- Technological Advances: Improvements in radar, imaging, and data analysis have increased both the number and quality of UAP reports. This fact renders them harder to dismiss as mere artifacts or errors. Even satellites can provide additional coverage while searching the skies for answers.

Congressional Hearings: Recent Developments and Future Hope

The 2022 and 2023 congressional hearings marked a turning point. Officials testified under oath about the number of sightings, the rigor of investigations, and the possible implications for national security. Notably, some witnesses hinted at the existence of materials or even biological evidence of non-human origin, though these claims remain unverified publicly.

Lawmakers on both sides of the aisle expressed bipartisan interest in continuing the investigation. Committees on intelligence, armed services, and science all called for further examination, the establishment of specialized offices to investigate UAPs, and mandatory reporting requirements for military and civilian pilots.

The witnesses and whistleblowers include Tim Gallaudet, Luiz Elizondo, Journalist Michael Shellenberger, and the previously mentioned Dave Fravor. Gallaudet is an esteemed retired rear admiral with the U.S. Navy. Elizondo was the former head of the key Defense Department's Advanced Aerospace Threat Identification Program (AATIP). Shellenberger had already written about compelling revelations involving a shadowy UAP program created in 2017. Their joint testimonies were quite compelling, as far as I was concerned.

In particular, however, I found the 2023 testimony of David Grusch has been especially controversial. Grusch is a former intelligence official whose explosive claims have energized both believers and skeptics. The detailed examination of Grusch's testimony before Congress is as fascinating as is the reactions it has provoked.

David Charles Grusch is a decorated former United States Air Force officer and a veteran of both the National Geospatial-Intelligence Agency (NGA) and the National Reconnaissance Office (NRO). During his tenure with these highly secretive agencies, Grusch reports that he served in roles that involved oversight of UAP-related analysis and investigations. He was a member of the U.S. government's Unidentified Aerial Phenomena Task Force (UAPTF), which was established to analyze reports of unexplained aerial sightings, particularly those encountered by military personnel.

In June of 2023, David Grusch stepped into the spotlight, both through media interviews and a formal appearance before the United States House Oversight Committee. His testimony marked one of the most significant moments in the modern history of UAP investigations, in my opinion. Grusch made several headline-grabbing claims, delivered under oath, that went far beyond what most other government officials had previously admitted regarding UAPs.

Key claims made by Grusch includes:

- Agencies of the U.S. government are in possession of non-human craft
- Agencies within the U.S. government have recovered and studied vehicles of non-human origin. He maintained that some of these craft were discovered to be wholly intact, while others were partially damaged.
- Biological evidence reportedly includes non-human intelligent beings, some of which may be working with governmental organizations that are working to reverse engineer their craft.

Considering what I myself saw in 1992, I would say that I believe Grusch's testimony to be true. Especially since I can personally confirm his testimony!

Legislative and Institutional Changes Related to these Hearings

Congress has begun to institutionalize the study of UAPs. The creation of the All-domain Anomaly Resolution Office (AARO), tasked with collecting, analyzing, and reporting on UAP data across all branches of government, marks a significant step. Legislation now compels agencies to produce regular reports, to catalog past incidents, and to ensure whistleblower protections for those with knowledge of secretive programs. However, AARO's reports and revelations have been seen by many, including me, as being completely underwhelming.

Furthermore, there are ongoing discussions, as was noted again in the 2024 congressional hearings. The subjects that were raised included increased funding for UAP investigations, facilitated cross-agency collaboration, and enhanced technological capabilities to better collect reliable data. These efforts aim to create a more comprehensive understanding of UAPs and ensure that no relevant information slips through bureaucratic cracks.

The Future Trajectory of Congressional Hearings

Looking ahead, there are several trends that are likely to shape the future of congressional hearings on UFOs:

1. Increased Frequency and Public Access

As public interest remains high and lawmakers recognize the political capital in addressing such a mysterious issue, Congress is likely to hold more frequent hearings, even though they cancelled the expected May of 2025 enquiries. These future sessions will be expected to feature not only defense officials but also a variety of scientists, civilian witnesses, and recognized independent experts.

2. Scientific Rigor and Collaboration

Future hearings will likely bring greater emphasis on scientific methodology. Lawmakers are expected to invite astrophysicists, engineers, and data analysts to help distinguish between natural phenomena, technological artifacts, and genuinely unexplained cases.

International collaboration may become a priority. UAPs are not a uniquely American occurrence, as I believe I have demonstrated. Countries like Brazil, France, Argentina, Britain, and Japan have their own investigative programs. Congressional hearings may lead to joint initiatives, data sharing, and the establishment of global standards for reporting and investigation.

3. Institutional Accountability and Oversight

With the proliferation of whistleblower reports and claims of hidden programs, Congress will continue to focus on accountability. Future legislation may create independent review boards or special inspectors general to audit UAP investigations. Additional protections for whistleblowers and requirements for timely, unfiltered briefings to oversight committees are likely to be discussed and enacted.

4. The Role of Emerging Technologies

Advancements in artificial intelligence, satellite surveillance, and progress in updated sensor networks, all promise to revolutionize the study of UAPs. Future hearings may explore how these technologies can be leveraged to provide real-time tracking, identification, and analysis. This certainly might include any of the breakthroughs of military contractors. We would certainly include Lockheed Martin into this group.

Congress may also probe the ethical and privacy implications of increased surveillance, ensuring that scientific curiosity and national security do not infringe on the rights of citizens.

Potential Challenges and Controversies

Despite increasing transparency, the future of congressional hearings on UFOs is not without a great deal of obstacles, as has been noted. All we must do on this front is ask: what happened to the May of 2025 hearings?

- Secrecy vs. Openness: Intelligence and defense agencies may resist full disclosure, citing concerns for national security, foreign espionage, or technological advantage. What is the fine line this obligation requires?

- Credibility of Witnesses: As more individuals come forward, the challenge will be to separate credible testimonies from speculation, misinterpretation, or hoaxes. We have had a combination of realistic observations and outright fakes.
  This is not unique to UAPs, we must note. Whenever a new whistleblower comes forward (and this includes me), their veracity will immediately be of concern to those on both sides of the argument.

- Political Polarization: As with many issues in Congress, there is a risk that UAP discussions could become politicized, undermining bipartisan cooperation and public trust. Nevertheless, it is fascinating how this topic appears to bridge the gap between the two parties. We can only hope this trend continues, and even grows.

- Public Expectations: The allure of sensational revelations may overshadow the often slow, methodical process of scientific inquiry. This could easily lead to disappointment or the promotion of further conspiracy theories if answers are not forthcoming.

Implications for Society and Science

The future of congressional hearings on UFOs extends far beyond the halls of government. If managed with integrity and rigor, these proceedings have the potential to:

- Enhance public trust in government transparency and accountability
- Promote interdisciplinary scientific research and technological innovation
- Encourage international cooperation on issues of shared interest
- Foster a culture of open inquiry and critical thinking
- Demystify unexplained phenomena, regardless of their ultimate origin

My Final Conclusions:

The future of American congressional hearings on UAPs is one of real promise, definite challenges, and the possibility for significant transformation. By embracing increased transparency, scientific rigor, and accountability, Congress can optimistically move the discussion beyond mere speculation and out from behind the veil of powerful governmental secrecy.

Whether the ultimate answers lie in some pure alien technology, advanced foreign technology, misunderstood natural phenomena, or something truly as extraordinary as the unique craft I saw on Cape Canaveral Air Force Station back in 1992, the challenging journey toward an accurate understanding of this topic may be as important as is the destination.

While the world earnestly watches, the interplay between our curiosity, persisting skepticism, and the possibility of great discovery will shape not just the future of UAP investigations, but also our collective approach into the unknown. Under the shadow of quantum computers, advancing artificial intelligence, and global internet interdependence, the people of our planet struggle to understand what our technology is doing to us, and for us. We hang in the balance of the world we once thought was much more straightforward and the growing mysteries of the universe we are seeking to learn and conquer.

Only time can show us what twists and turns lie before us as Americans, world citizens, and as the human species!

About the Author

Dr. Gregory Rogers has served as a flight surgeon within the Department of Defense since 1984. He was initially stationed in West Germany during the Cold War days. As part of his duties, he assisted during the deployment of the nuclear Pershing II inter-regional ballistic missiles (IRBMs) to West Germany. He was solo-certified in the TH-55, but also flew the UH-1 Huey, the OH-58 Scout, and the AH-1 Cobra attack helicopter.

Dr. Rogers was next assigned to the 45th Space Wing at Patrick Air Force Base, Cape Canaveral Air Force Station, the Eastern Space and Missile Center, and the Eastern Missile Range. He performed air-sea rescue missions with the 41st Air Rescue Squadron and the 301st Rescue Squadron. He was one of the first flight surgeons who deployed to South Florida after the devastation that was wrought by Hurricane Andrew. He also supported the Air Force Technical Application Center (AFTAC), the Joint STARS development test wing, and elements of the 9th Strategic Reconnaissance Wing.

Being attached to NASA for support of the manned spaceflight program, he supported 31 space shuttle launches, 14 landings at Kennedy Space Center, and numerous rescue and contingency exercises at KSC. He authored the Transoceanic Abort Landing summary, and the Kennedy Space Center Shuttle Landing Facility support documents for Department of Defense flight surgeons for their possible operations to rescue and treat the NASA astronauts.

Dr. Rogers was also trained and flew the T-37 jet trainer, the supersonic T-38 Talon and the F-16 Fighting Falcon, also known by its pilots as the

Viper. He was awarded with two Sikorski Rescue Awards, the Humanitarian Service Medal, the Air Medal, and the Meritorious Service Medal.

Dr. Rogers and his wife currently live in McAlester, Oklahoma

Publications by Un-X Media

Rules for Goddesses by Margie Kay 1999
Haunted Independence Missouri by Margie Kay 2013 & 2016
Gateway to the Dead: A Ghost Hunter's Field Guide
by Margie Kay 2016
Family Secrets by Jean Walker 2017
The Kansas City UFO Flaps by Margie Kay 2017-2025
Un-X News Magazine 2011-2024 in print and digital
A Sonoma County Phenomenon by Margie Kay 2019
The Remote-Viewing Workbook by Margie Kay 2019 (on LULU)
The Fast Movers: Evidence for High-Speed UFOs/UAPs
by Margie Kay, Bill Spicer, and Larry Tyree 2020
Journey to Spirit by Devin Listrom 2020
Winged Aliens by Margie Kay 2021
The Master Dowsers Chart Book by Margie Kay 2021 (on LULU)
The Alien Colonization of Earth's Waterways
by Debbie Ziegelmeyer 2021
50th Anniversary of the SE Missouri Ozarks UFO Flap
by Debbie Ziegelmeyer and Margie Kay 2022
Meeting Wallace by Larry Costa 2023
Poems by Pat Delap by Pat Delap 2024
Holiday Poems and Recipes by James Bair 2023
Dying to Meet Them by Mindy Tautfest 2024
All Monsters are Human by Derrick Smith 2025
Missouri: UFO Hot Spot by Missouri MUFON 2025
THOR: The Extraterrestrial on Earth by Margie Kay 2025
Incident in Varginha by Vittorio Pacaccini and Fernanda Pires 2024
UFO Attacks in Brazil by Thiago Luiz Tichetti 2025
We Got it From THEM by Dr. Gregory Rogers

Adult Coloring books

Documentary films:
PORTALS
Mysterious Missouri
THOR

Most books are available at www.amazon.com

Un-X Media is currently taking book submissions. We publish books about unexplained phenomena. Please check the website for writer guidelines. www.unxnmedia.com